# BLOCKCHAIN TECHNOLOGY FOR PAPERLESS TRADE FACILITATION IN MALDIVES

DECEMBER 2020

ASIAN DEVELOPMENT BANK

ADB

© 2020 Asian Development Bank
6 ADB Avenue, Mandaluyong City, 1550 Metro Manila, Philippines
Tel +63 2 8632 4444; Fax +63 2 8636 2444
www.adb.org

Some rights reserved. Published in 2020.

ISBN 978-92-9262-605-1 (print); 978-92-9262-606-8 (electronic); 978-92-9262-607-5 (ebook)
Publication Stock No. TCS200403-2
DOI: http://dx.doi.org/10.22617/TCS200403-2

Notes:
In this publication, "$" refers to United States dollars.

Cover design by Francis Joseph M. Manio.

Printed on recycled paper

# Contents

# Tables, Figures, and Box

# Acknowledgments

This report aims to understand the trade ecosystem in Maldives and select a relevant pilot use case to demonstrate the significance and relevance of blockchain technology in international trade.

The findings in this report are based on information gathered during workshops conducted in Malé from 25 March to 4 April 2019. Key stakeholders from Maldives customs, ports, and trading organizations were invited to face-to-face meetings and group workshops during the aforementioned period. During such discussions, bottlenecks in current trade processes and technology infrastructure were discussed, in detail.

Based on current and future requirements of the country, the list of blockchain use cases were identified. The use cases were then evaluated on a set of qualitative and quantitative parameters to arrive at the most suitable use case for a pilot deployment.

The pilot use case was selected to demonstrate the ability of blockchain technology to improve the overall import trade process at Maldives. Key performance indicators after successful implementation of the pilot would be digitization of trade documents, seamless payments among participants, and marked improvement in time to process trade requests.

Kartik Burman, international consultant for the Asian Development Bank's (ADB) South Asia Subregional Economic Cooperation National Single Window Project, prepared this report, supported by IBM India Private Limited.

Takeru Toba, economist, ADB, and Masato Nakane, senior economist, ADB, oversaw the preparation of this study. Seok Yong Yoon, principal public management specialist, ADB, provided technical guidance, and Ivan Magadia, consultant, ADB, assisted in the production process and provided valuable input and feedback in finalizing the report.

The project team is thankful to the following organizations for the successful conduct of trade process discussion in Maldives: Airport Ferry Operators Association of Maldives, Allied Insurance Company of the Maldives Private Limited, Asia Forwarding Private Limited, Bank of Maldives PLC, Bonito Cargo and Logistics Private Limited, Centurion PLC, Lily Shipping Private Limited, Maldives Airports Company Limited, Maldives Customs Service, Maldives Food and Drug Authority, Maldives Islamic Bank Private Limited, Maldives Ports Limited, Ministry of Economic Development, Ministry of Defence and National Security, Ministry of Fisheries, Marine Resources and Agriculture, Maldives Inland Revenue Authority, Ministry of Finance, Ministry of Communication, Science and Technology, Prime Fertilizers Private Limited, and State Trading Organization PLC.

# Abbreviations

| | |
|---|---|
| API | application programming interface |
| ASYCUDA | Automated System for Customs Data |
| CBRA | cross-border regulatory agency |
| COSYS | cargo operations system |
| EDIFACT | electronic data interchange for administration, commerce, and transport |
| ETA | estimated time of arrival |
| MACL | Maldives Airports Company Limited |
| MCS | Maldives Customs Service |
| MED | Ministry of Economic Development |
| MFDA | Maldives Food and Drug Authority |
| MFMRA | Ministry of Fisheries, Marine Resources and Agriculture |
| MIRA | Maldives Inland Revenue Authority |
| MPL | Maldives Ports Limited |
| NSW | national single window |
| SOAP | simple object access protocol |

# Executive Summary

In today's global economy, cross-border trade is fundamental to economic activity and growth. In cross-border trade, goods and accompanying documents are exchanged through an intricate network of trade partners, ports and regulatory authorities.

The efficiency with which goods and documents are exchanged, across multiple parties, both domestic and international, determines the overall effectiveness of a trade network. Removing friction, building trust, and improving transparency are the key requirements for improving any trade network.

In this report, key challenges related to the current import trade process at Maldives are discussed, followed by understanding how blockchain can address these challenges, more effectively, than traditional technologies and finally selecting a suitable use case for pilot testing.

A description of technical architecture for the blockchain pilot is provided in section C.2. The key theme includes building a pilot application that can be scaled to a production system with minimal effort and allow easy onboarding (adding new participants, like cross-border customs, ports, banks), in the future.

The current trade process in Maldives is challenged by various issues like prevalence of fraudulent trade documents, repetitive manual processes, and poor availability and usability of key trade information technology systems. The fragmented and nonstandardized way of developing the national digital infrastructure is also posing a challenge to build a single trade platform for the country.

While traditional approach and technologies can address these challenges, blockchain technology, by offering a single portal solution, offers a network solution that is built to integrate seamlessly with both domestic and cross-border systems and processes.

Blockchain technology offers immutable records of transactions and will bring required levels of trust and transparency in the trade ecosystem. Data once committed to the blockchain ledger cannot be altered, at any point in time, and can be considered as golden source of truth, where multiple parties have different views, as per current process.

Furthermore, transfer of ownership for key trade documents is fundamental to the workings of international trade. Many of these document ownership transfers are also linked with payments. Blockchain smart contracts coupled with immutable ledger system can securely and transparently manage these ownership transfers and facilitate online digital payments more efficiently.

Most importantly, the success of any new trade system heavily relies on its ability to scale in the future and allow for easy onboarding of new and cross-border participants. Blockchain as a technology with distributed node architecture is built to scale efficiently and allows for easy integration with different systems, at varied levels of maturity.

Out of the six use cases identified during the study, one was selected, based on strategic importance and feasibility of deployment. The use case selected would offer a single blockchain platform to domestic and cross-border participants, to share key trade documents, make online payments and transparent exchange of data and document ownerships.

## A.1 International Trade Industry Circle

The international trade process involves multiple government and private parties. These network participants (or network members) interact with each other to exchange documents, make or receive payments, enter or amend data, and issue approvals or denials.

The health of such trade network depends on trust, accuracy, and speed at which data and documents can be exchanged and processed, among the network members. Fragmented processes and technologies could pose a

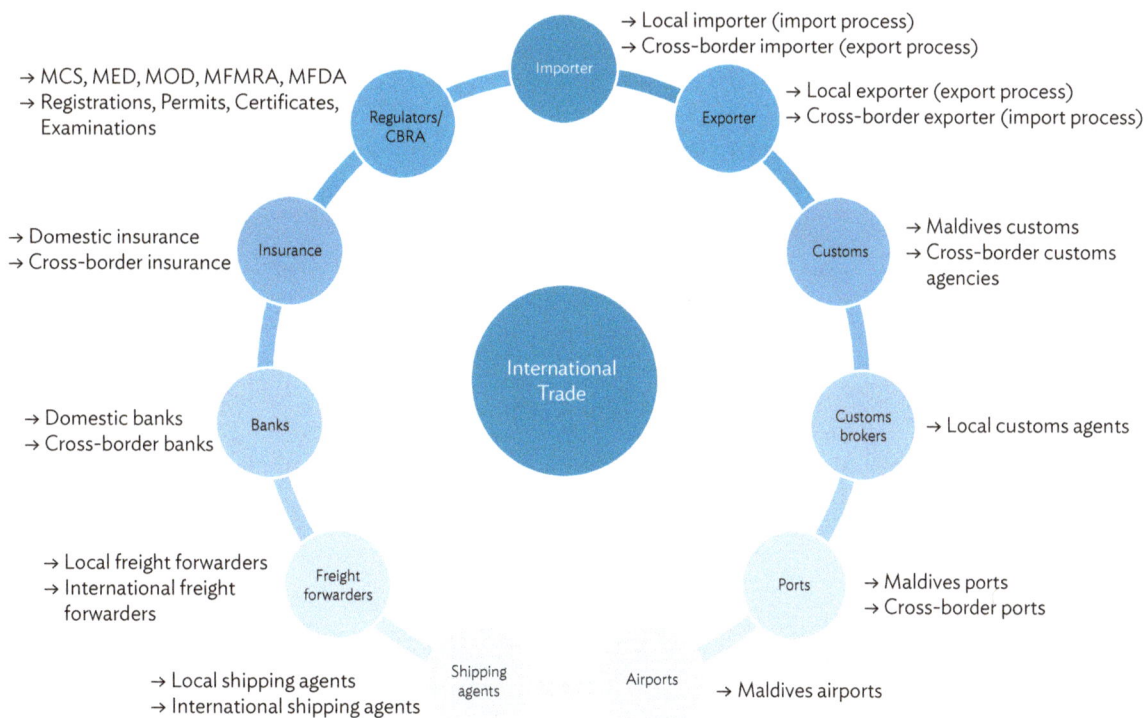

Figure 1: International Trade Industry Circle

→ Local importer (import process)
→ Cross-border importer (export process)

→ Local exporter (export process)
→ Cross-border exporter (import process)

→ MCS, MED, MOD, MFMRA, MFDA
→ Registrations, Permits, Certificates, Examinations

Importer

Exporter

Regulators/CBRA

Customs
→ Maldives customs
→ Cross-border customs agencies

→ Domestic insurance
→ Cross-border insurance

Insurance

International Trade

Customs brokers
→ Local customs agents

→ Domestic banks
→ Cross-border banks

Banks

→ Local freight forwarders
→ International freight forwarders

Freight forwarders

Ports
→ Maldives ports
→ Cross-border ports

→ Local shipping agents
→ International shipping agents

Shipping agents

Airports

→ Maldives airports

CBRA = cross-border regulatory agency; MCS = Maldives Customs Service; MED = Ministry of Economic Development; MFDA= Maldives Food and Drug Authority; MFMRA= Ministry of Fisheries, Marine Resources and Agriculture; MOD= Ministry of Defence.

Source: Author's analysis.

barrier to the seamless exchange of information. Operational procedures that are opaque to regulators increase the risk of fraud and tax evasion.

Furthermore, a trade system that keeps in view only short-term domestic requirements for data and documents processing—while limiting cross-border interactions from a data security, confidentiality of proprietary business rules, or network governance perspective—is not a trade system built for the future. The future is about collaboration and cooperation, which will lead to more trust, transparency, and collaboration.

Figure 1 depicts the international trade circle for Maldives. Interactions among these network members throughout the trade life cycle form the basis for this report's approach to understand the key challenges, requirements, and opportunities for technology intervention.

# A.2 Key Pain Points in International Trade

Pain points refer to the challenges faced by trade participants in Maldives. These are related to current processes and technologies that pose a barrier to implement a transparent, efficient, and trustworthy trade ecosystem.

Table 1 provides the list of key pain points identified by the study team. These are based on responses to questionnaires, in-person workshops, and focus group discussions.

The focus group consists of government and private organizations from Maldives, which are directly involved in the cross-border trade process.

### Table 1: Key Pain Points in International Trade

| No. | Pain Points | Importer | Exporter | Customs | Customs Broker | Ports | Airports | Shipping Agent | Freight Forwarders | Banks | Insurance | Regulators |
|---|---|---|---|---|---|---|---|---|---|---|---|---|
| 1 | Risk of trade document fraud | ● | | ● | | | | | | | | ● |
| 2 | Multiple accounts lock-in | ● | ● | | ● | | | ● | | | | |
| 3 | Process duplication | ● | ● | ● | ● | ● | ● | ● | ● | | | ● |
| 4 | Compliance enforcement | | | | ● | | | | | | | ● |
| 5 | Cross-border collaboration | | | ● | | ● | ● | | | ● | ● | |
| 6 | Trust and reliability | ● | ● | ● | ● | ● | ● | ● | ● | ● | ● | ● |
| 7 | Standardization | ● | ● | ● | ● | ● | ● | ● | ● | ● | ● | ● |
| 8 | Security and privacy | | | ● | ● | | | | | ● | ● | ● |
| 9 | Process transparency | ● | ● | ● | ● | ● | ● | ● | ● | ● | ● | ● |
| 10 | Manual payments | ● | ● | | ● | ● | ● | ● | | ● | | ● |

Source: Author's analysis.

## Risk of Trade Document Fraud

In international trade, risk of counterfeit or false documents is a major concern. Maldives Customs Service (MCS) highlighted the issue of document fraud, during import declaration process, leading to errors in price determination and revenue loss for the government.

## Multiple Accounts Lock-in

Various trade partners like customs brokers, shipping agents, trade organizations, and consignees have to maintain multiple pre-funded accounts with service providers and government agencies. Individual accounts maintained with MCS, Maldives Ports Limited (MPL), Maldives Airports Company Limited (MACL), and cross-border regulatory agencies (CBRAs) lead to working capital lock-in and inefficient utilization of available funds.

## Redundant Document Exchange

There is no single platform for document submission. Trade documents are submitted multiple times to multiple agencies, and sometimes as physical hard copies. This leads to repetition, disputes, and time delays.

## Compliance Enforcement

The lack of intelligent and integrated systems to enforce compliance is a major challenge. Delays in document submission are penalized with a fine, but this is imposed through manual channel, causing delays and errors. Compliance requirements are followed as reactive procedures and not enforced as system or process-level standards.

## Cross-Border Collaboration

The unavailability of digital channels for cross-border collaboration between domestic and overseas customs or ports is a key pain point. According to MCS, a customs-to-customs collaboration platform would vastly improve trust and transparency in the cross-border documentary exchange process.

## Trust and Reliability

As trade documents are submitted multiple times and manually to different government offices, the risk of fraudulent document exchange and data verification oversight is quite high. To mitigate this risk, each document is reviewed multiple times, further extending the long processing times.

## Standardization

The lack of document and data standardization was highlighted by multiple agencies as a key concern for collaboration and integration. While implementation of Automated System for Customs Data (ASYCUDA) World, **Maldives Customs technology platform for processing trade-related data,** has brought in some data standardization in MCS, data harmonization among other agencies is a key pain point.

## Security and Privacy

Standards related to data security and privacy are under development in Maldives. There is also concern about poor infrastructure for data storage and recovery. All key government departments have a single point of failure, and primary and backup databases are physically collocated.

## Process Transparency

The poor process-level transparency across the trade life cycle is attributed to limited system integration between agencies and manual interventions. For example, the consignee does not have insight into the trade document processing status. MPL, which relies heavily on the availability of critical documents like estimated time of arrival (ETA), does not have a real-time view into such documents.

## Manual Payments

Most payments are done manually using cash or check. Pre-funded accounts are also used for speedy clearance of goods, but this leads to inefficient use of working capital. The primary reason behind the use of manual payment methods is attributed to higher transaction costs for online payments.

Manual payments lead to longer processing times, additional fixed costs for dedicated staff, a lengthy transaction cycle for dispute resolution, and additional steps for processing rejected checks.

# A.3 Current Import Trade Process

The current trade process in Maldives is divided into five segments:

    (i)    permits and registrations,
    (ii)   import declaration,
    (iii)  shipping documents exchange,
    (iv)  ports and air cargo operations, and
    (v)   goods examination and clearance by CBRAs.

Each segment has multiple points of interactions among various trade participants and is dominated by manual exchange of data and documents.

Key information technology applications associated with the import trade operations are ASYCUDA World, used by MCS; Bandharu, used for port operations by MPL; and cargo operations system (COSYS), used for air cargo operations by MACL.

## A.3.1 Permits and Registrations

### *Customs Registration*

The customs registration process starts with either manual (in-person submission of an application form at the MCS office) or digital submission of an application form through the Maldives Customs Portal Plus. The process, depicted in Figure A.1, performs an automatic check for Ministry of Economic Development (MED) business registration data, through an application programming interface (API) connection between the MCS and MED e-portals. A successful application process leads to the generation of a customs registration number (C series for consignees and B series for customs brokers).

All customs data are stored at the primary data center, physically located in the MCS office building. The backup data are also maintained in the same building.

### Business Registration

The business registration process is achieved by manual submission of an application to the MED office, or by digital submission through the MED e-portal. The process is depicted in Figure A.2. The MED system (BOLI) is integrated with the Maldives Inland Revenue Authority (MIRA) tax database through an API connector. The business registration process with the MED is complete when the internal checks, along with the availability of tax identification number, are confirmed with MIRA. The business registration number is mandatory to apply for customs registration and permits from CBRAs.

### Permits Process

For import or export of goods, the consignee also needs to request relevant permits with CBRAs. This is applicable for each consignment. While the MED and the Ministry of Defence have developed an e-portal to facilitate the permit application process, other departments like Maldives Food and Drug Authority (MFDA) and the Ministry of Fisheries, Marine Resources and Agriculture (MFMRA) only allow manual submission of applications. The process is depicted in Figure A.3.

The import trade process includes multiple redundant interactions among CBRAs, especially between the MED, MFDA, and MFMRA. The applicant has to get physical signatures from respective departments on the permit application form, which is time-consuming, repetitive, and cumbersome.

### Maldives Identity Management System

The Government of Maldives has rolled out a digital identity service for its citizens and businesses, called eFaas. The service allows various government agencies to streamline authentication services and allow faster access to public services. The eFaas application uses the OpenID Connect protocol to allow client applications to verify the identities of end users. eFaas creates an online identity for citizens, residents, and organizations based on government-issued documents. Its functionality will be extended in the future to use biometric and demographic data.

## A.3.2 Import Declaration

### Customs Declaration

The customs declaration process is depicted in Figure A.4. The customs broker can submit the import declaration form online, using the ASYCUDA World gateway, or manually at the customs office, at Malé. All declaration forms are subjected to a check by MIRA to assess outstanding tax. MIRA's check will be carried out automatically through the web-service layer already developed between ASYCUDA World and MIRA systems.

ASYCUDA World will generate the assessment number and payment instruction once the internal review and risk assessments are complete. The payment details will be mapped to the assessment number and will be made available for online payment on the e-customs portal. The consignee or customs broker can make the payment online through the Bank of Maldives payment gateway or manually at the customs office in Malé.

All payments made toward import duty are deposited to a Ministry of Finance account maintained at Maldives' central bank, Maldives Monetary Authority. Cash and check payments are manually deposited the next day at the Maldives Monetary Authority office. In case of check returns, the return notice is sent manually to the customs broker at the customs office in Malé.

ASYCUDA World is a thick client[1] application that needs to be installed locally by the customs brokers, shipping agents, freight forwarders, and any other third-party agents. This is a resource-intensive application and does not support bulk upload of files (XML files for declarations). This creates operational inefficiencies in the document exchange process.

The custom brokers have to maintain a pre-funded account with MCS for faster processing of declarations and amendments. This leads to liquidity lock-in with MCS and inefficient use of capital by brokers.

## A.3.3 Shipping Documents Exchange

### *Cargo Manifest*

As part of the shipping documents exchange process, the customs broker uploads the original bill of lading at the time of import declaration, and the freight forwarder should submit the second carrier bill of lading before the arrival of the shipment. This file can be submitted as an XML upload to ASYCUDA World. Based on data available in ASYCUDA World, the customs agent will manually check whether the shipment has arrived and whether the manifest has been submitted by the shipping agent. In case the cargo has already arrived, and the bill of lading entry is not available in ASYCUDA World, a fine will be imposed on the shipping agent.

The shipping agent also needs to submit the ETA document, in hard copy, for the incoming shipment to MCS and MPL.

The customs agent manually checks for any manifest or bill of lading amendments, and issues payment notification accordingly. This is followed by a manifest write-off after matching the manifest with the bill of lading. The process is depicted in Figure A.5.

In case the bill of lading entry is not available, and the shipment has already arrived, a notice of fine will be issued to the shipping agent. For any amendments to the manifest document that may arise due to the second carrier bill of lading, a payment notice will be sent to the customs broker.

MCS (through ASYCUDA World) will also receive the air waybill for incoming cargo through the COSYS integration layer (via an API call). COSYS is the IT system managed by MACL.

---

[1] A thick client is a computer that does not require a connection to a server system to run (unlike a thin client). However, they can benefit from connecting to a network and a server.

## A.3.4  Ports and Air Cargo Operations

### *Maldives Ports Limited Operations: Import Trade*

MPL operations start with the receipt of the ETA document from the shipping agent. Details of the ETA document are then manually entered into the Bandharu, the port management IT system. The process is depicted in Figure A.6.

Per the current process, the freight forwarder submits the original bill of lading to the shipping agent. The shipping agent creates the manifest document in its internal system. The shipping agent then submits the original bill of lading to MPL, both as a physical copy and an XML file via email. In case only a physical copy of the bill of lading is received, the MPL data entry operator enters the bill of lading details into Bandharu.

Before the arrival of the ship, the shipping agent also has to submit the cargo manifest, packing list, container list, and vessel plan to the cargo department of MPL. Data fields from these documents are manually entered into Bandharu.

MPL requires the submission of the original or second carrier bill of lading, 24 hours before the arrival of the vessel, but in many cases these documents are not submitted as per the designated timeline.

At the time of vessel arrival, the shipping agent can raise the request for various port services like pilotage through the online portal of MPL, or by submission of a physical application form. These entries are updated in Bandharu by the MPL data entry staff. Although MPL facilitates online service charge payments, 99% of shipping agents actually make these payments either by cash or check because of the high transaction processing charges imposed by MPL for online payments.

In case of shipment involving containers, the shipping agent also needs to submit the inward list for container unloading. This list can be sent as an XML file via email to MPL. These data are then manually entered in Bandharu.

Once the shipping agent issues the delivery order to the consignee, the consignee can approach MPL with original copies of the delivery order and bill of lading for clearance. After a manual check of all clearance documents, including the delivery order, the bill of lading is completed, and the delivery process will be initiated. At this point, an invoice will be issued to the consignee for port charges like storage charges (more than 10 days at a storage location), power consumption charges, express service charges, re-measurement charges, handling charges and shifting charges.

After payment for port charges are cleared, the consignee can raise the shifting request with MPL to move the goods to the examination area. Once the goods are examined and released by the concerned CBRAs, the release order is issued for the goods approved for release. The gate pass is a physical document that is signed and stamped by MPL and the concerned CBRAs.

Most processes at MPL are manual, with limited integration with other trade-related systems.

### Air Cargo: Import and Export Operations

The air import and export processes at MACL are largely managed by the COSYS system. This system has interfacing capabilities with airline systems (to exchange Société Internationale de Télécommunications Aéronautiques messages) and the ASYCUDA World system with MCS (Figure A.7).

Before the arrival of the consignment, the trade documents (e.g., flight-forwarding messages and air waybills) are received as Société Internationale de Télécommunications Aéronautiques messages by COSYS and shared with ASYCUDA World, via an API call.

The declaration process is initiated by the customs broker and follows a similar assessment procedure  as that of sea import section A.3.2). The payments for MACL charges are collected by cash or by check. Customs brokers generally maintain a pre-funded account with MACL, in addition to similar accounts with MPL and MCS, to be used toward payments of service charges.

As most of the import and export cargo from Maldives airports consists of perishable goods, faster movement of consignments through the administrative process is considered critical.

### Export Trade

The export trade process is less complex than the import process. Figure A.8 depicts the relatively simple interlinked process flow.

The customs broker plays a pivotal role in the export document exchange, port service request, and payments toward customs and port services. As fisheries products constitute the major part of the country's export business, the MFMRA and MFDA play a crucial role in ensuring the required certificates and quality standards are enforced.

Once the export declaration form is received, either manually or through ASYCUDA World, MCS performs the internal checks, which include verification of relevant permits, business registration data. Once the internal assessment process is complete, MCS issues the assessment number. The customs broker is informed about the duties due, which can be paid to MCS (by check) or via online payment using the e-customs link to the Bank of Maldives payment gateway.

The MFMRA collects payments toward the processing and issue of catch certificates, while MFDA collects payments toward the processing and issue of health certificates. The current process for issuing these certificates is manual in nature. Before the goods are released, final checks are completed by the CBRAs. A paper document trail is maintained among the CBRAs to complete the checks and stamping process.

## A.3.5  Goods Examination and Clearance by Cross-Border Regulatory Agencies

### Customs Clearance

Controlled goods are moved to a bonded warehouse for inspection, while general goods are inspected at the port. The process is depicted in Figure A.9.

Per the existing process, MCS agents have to wait for the broker or the consignee to physically approach them at the inspection area. No notification systems are available to MCS to inform the concerned party (i.e., the consignee or customs broker) about the timing or requirements for such examination.

While MCS has a tablet version of the e-customs application for goods inspection, the decision to inspect the entire cargo or a selection of the incoming goods is manual in nature and not aided by any IT systems.

One of the major challenges for MCS is to correctly determine the price of the item imported into the country. There have been incidents when the price declared on the invoice or the purchase order does not match the printed price on the goods. While this discrepancy can only be determined during physical inspection of goods, this leads to inefficient use of time and limited inspection space.

During inspection, goods are checked against import declarations. If all data check points match, then goods will be marked for release. If not, the following actions are taken:

(i)     In case of an excess, the consignee will be charged a fine and asked to resubmit the declaration with amendments or submit a new declaration. In case of a shortfall, MCS will modify the declaration and a refund of duty will be issued.

(ii)    In case of partial goods arrival, the existing declaration will be updated for reuse during the next consignment for leftover goods to be received in upcoming cargo.

# A.4 Current Technology Architecture for Trade in Maldives

Figure 2 depicts the technology architecture for import trade applications in Maldives.

In summary, the import trade application landscape is heavily dominated by thick client-side applications, integrations over simple object access protocol (SOAP), and manual uploads of XML files.

The mainstay system for this landscape is ASYCUDA World. It is a thick client application for users to prepare and upload declaration, manifests and supporting documents. User access to the system is managed by database credentials, which is prone to security vulnerabilities.

Most of the applications are developed on C# or Java with back-end databases on Microsoft SQL. Microsoft SQL is not an optimal solution for handling database create, read, update, and delete operations. Heavy loads on such systems lead to severe throttling of responses and poor availability of resources.

Most of the integrations are through SOAP. SOAP is a legacy integration pattern and is bulky because of XML interaction. SOAP negatively impacts the overall performance of integrated applications.

In the current landscape, most of the applications do not have a data purging policy, including application logs data that are kept online. This makes the applications cumbersome, directly impacting file input and output operations.

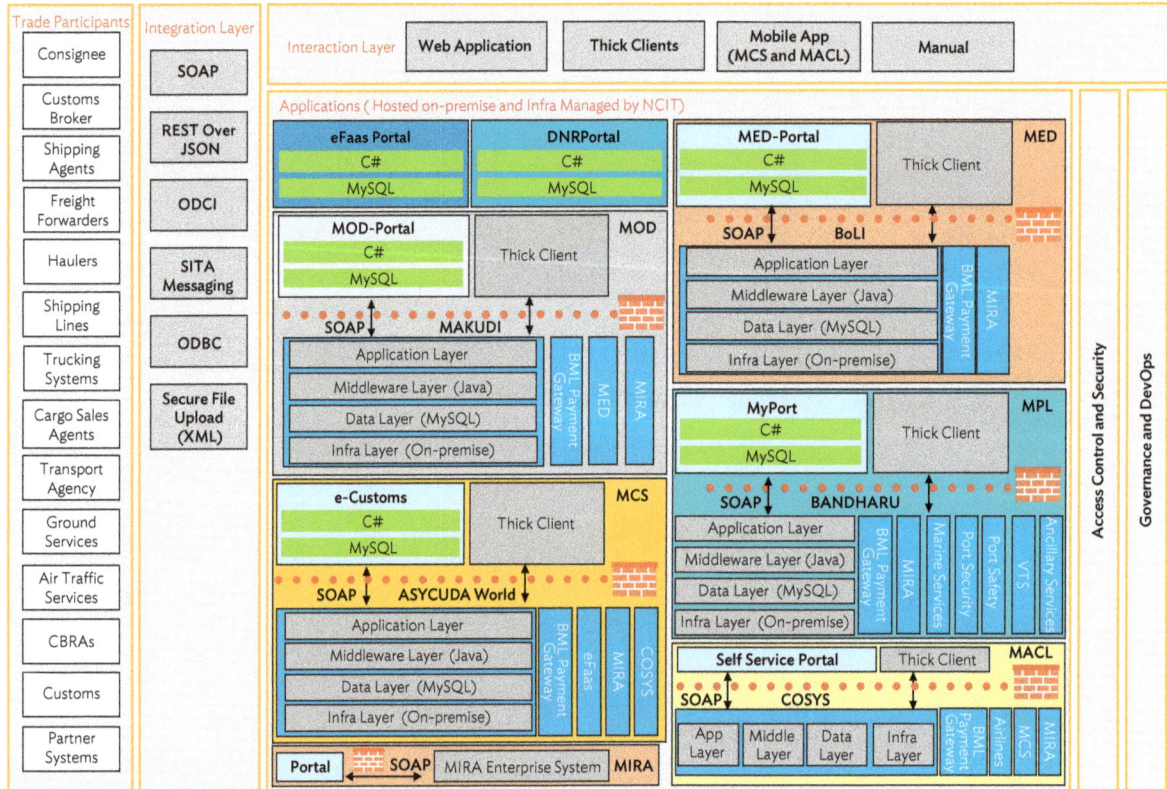

Figure 2: Current Technology Architecture for Import Trade Applications

ASYCUDA = Automated System for Customs Data, BML = Bank of Maldives Limited, CBRA = cross-border regulatory agency, COSYS= cargo operations system, DNR = Department of National Registration, JSON = JavaScript Object Notation, MACL = Maldives Airports Company Limited, MCS = Maldives Customs Service, MED = Ministry of Economic Development, MIRA = Maldives Inland Revenue Agency, MNDF = Maldives National Defence Force , MOD= Ministry of Defence, MPL = Maldives Ports Limited, NCIT = National Centre for Information Technology, ODBC = open database connectivity, ODCI = Oracle Data Cartridge Interface, REST = representational state transfer, SITA = Société Internationale de Télécommunications Aéronautiques, SOAP = simple object access protocol, VTS = Vessel Traffic Service.

Notes:
1.  MAKUDI refers to the IT system of MOD.
2.  eFaas refers to the Maldives Citizen Registration System.
3.  BOLI is name of the online MED portal.

Source: Author's analysis.

In terms of document management, sensitive trade-related documents and payment receipts are scanned and stored on a file system hosted and managed within an organization's IT infrastructure (applies to all trade-related organizations at Maldives). With poor version control practices and manual processing of data, the overall durability of the file system architecture is quite low.

Various IT stakeholders in the country have stressed the importance of a distributed data model to support an application failover scenario. The desired future model should also have applications with a data access layer for segregation of responsibilities between the database and applications. This will lead to improved database management capabilities.

In terms of infrastructure security, the National Centre for Information Technology provides a backbone for hosting of all e-government services (e-government), including applications related to trade business functions. This common infrastructure provision provides common firewall and application-level security protocols.

# Part B
# Blockchain Technology for Import Trade Process

Blockchain technology is associated with the finance sector, but it can be applied to other industries. The supporting architecture and features of blockchain presents tremendous opportunities to transform the trade ecosystem, as shown in the box.

## Box: Key Elements of Blockchain Technology

Blockchain is a shared, immutable ledger that facilitates the process of recording transactions and tracking assets in a business network. An asset could be either a physical entity or a digital record of value.

In the context of international trade, a blockchain network can be used to exchange data and documents as digital assets, or it can also be used to exchange value such as duty payments, fees, and charges.

Blockchain improves network efficiency by reducing duplication of work, automating network rules, and validating smart contracts. Once added to the blockchain network, data cannot be altered. This makes any blockchain solution highly trustworthy and removes friction among network members.

Key elements of blockchain:

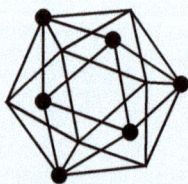

### Distributed Ledger Technology

All network participants have access to the distributed ledger and its immutable record of transactions. With this shared ledger, transactions are recorded only once, eliminating the duplication of effort that is typical of traditional business networks.

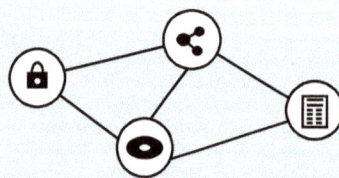

### Records Are Immutable

No participant can change or tamper with a transaction after it has been recorded to the shared ledger. If a transaction record includes an error, a new transaction must be added to reverse the error, and both transactions are then visible.

### Smart Contract

To speed transactions, a set of rules — called a smart contract — is stored on the blockchain and executed automatically. A smart contract can define conditions for corporate bond transfers, include terms for travel insurance to be paid, and much more.

Source: IBM. What is blockchain technology? https://www.ibm.com/my-en/blockchain/what-is-blockchain.

# B.1 International Trade Requirements: Blockchain and Non-Blockchain

Table 2 provides the list of requirements for an integrated cross-border trade system for Maldives. These requirements were identified during multiple workshops and one-on-one stakeholder meetings.

These requirements are then mapped to the relevant network participants. Some requirements, like single view of document, data ownership and control, provenance of goods, automatic reconciliation of records, and data security and privacy, affect all the participants in the network. Other requirements, like working capital management, affect only the participants maintaining individual pre-funded accounts with public and private agencies.

## Single View of Documents

Single view of trade documents is a key requirement for improving trust and transparency in the international trade process. Any participant logging into the trade platform will get the same view (version, updates, and history) of the document as any other member of the network. This functionality can be achieved using both blockchain and traditional technologies. However, blockchain technology offers this functionality as a default. This will result in faster development, testing, and deployment or upgrade cycles.

## Data Ownership and Control

While traditional technologies allow data and document sharing functionalities, when using a common middle layer, control over data is lost the very moment it is shared with the network. Also, in traditional architectures, proprietary business rules related to pricing and risk management need to be implemented on the common middle layer. Such architecture leads to friction in the network. Network members like customs, border control, and regulatory agencies may reject this idea of exposing internal business rules or their competitive position or proprietary operating procedures on a common network. The requirement to share data and documents on a common platform, and the ability to maintain control over this, is essential to reduce friction among network members.

Blockchain, with a built-in channel structure, allows network members absolute control over shared data. At the same time, individual organizations can keep their proprietary business rules secure by using smart contracts. Such capabilities of a blockchain system enable different countries to manage or maintain their business rules internally (within their geographical and operational boundaries) while still participating on an international trade platform.

## Single Workflow

Single workflow refers to a common network workflow that is not point-to-point between network members but instead encompasses all network members. This is essential to remove operational boundaries for business processes that transcend organizational or national limits.

While traditional technologies can be used to build workflows that interact with different organizations, the fragmented nature of such a solution defeats the primary objective of improved operational efficiency. On the other hand, a blockchain solution allows for a single workflow, with seamless integration end points across

## Table 2: Requirements for Maldives Trade Ecosystem Partners

| No. | Requirements | Importer | Exporter | Customs | Customs Broker | Ports | Airports | Shipping Agent | Freight Forwarders | Banks | Insurance | Regulators | Blockchain | Non-Blockchain (Traditional) | Time to Market |
|---|---|---|---|---|---|---|---|---|---|---|---|---|---|---|---|
| 1 | Single view of documents | • | • | • | • | • | • | • | • | • | • | • | ✓ | ✓ | Blockchain> Non-Blockchain |
| 2 | Data ownership and control | • | • | • | • | • | • | • | • | • | • | • | ✓ | ✓ | Blockchain> Non-Blockchain |
| 3 | Single workflow | • | • | • | • | • | • | • | • | • | • | • | ✓ | ✓ | Blockchain> Non-Blockchain |
| 4 | Smart contracts for network business rules | • | • | • | • | • | • | • | • | • | • | • | ✓ | | Blockchain>> Non-Blockchain |
| 5 | Built-in encryption and privacy | • | • | • | • | • | • | • | • | • | • | • | ✓ | | Blockchain>> Non-Blockchain |
| 6 | Transaction processing | • | • | • | • | • | • | • | | • | • | • | ✓ | ✓ | Blockchain>> Non-Blockchain |
| 7 | Immutability | • | • | • | • | • | • | | • | • | • | • | ✓ | | Blockchain>> Non-Blockchain |
| 8 | Network automated workflow | • | • | • | • | • | • | • | • | • | • | • | ✓ | ✓ | Blockchain>> Non-Blockchain |
| 9 | Cross-country trade process | • | • | • | | • | • | | | • | • | • | ✓ | ✓ | Blockchain>> Non-Blockchain |
| 10 | Data validation | • | • | • | • | • | • | | | • | • | • | ✓ | ✓ | Blockchain>> Non-Blockchain |
| 11 | Improved trust in the network | • | • | • | • | • | • | • | • | • | • | • | ✓ | | Blockchain>> Non-Blockchain |
| 12 | Built-in analytics and cognitive | • | • | • | • | • | • | | | • | • | • | ✓ | | Blockchain>> Non-Blockchain |
| 13 | Working capital management | • | • | | • | | | • | | | | | ✓ | | Blockchain>> Non-Blockchain |
| 14 | Near real-time payments | • | • | • | • | | • | • | | • | • | • | ✓ | | Blockchain>> Non-Blockchain |
| 15 | Compliance enforcement | | | • | | | | | | | | • | ✓ | ✓ | Blockchain>> Non-Blockchain |

*continued on next page*

*Table 2 continued*

| No. | Requirements | Importer | Exporter | Customs | Customs Broker | Ports | Airports | Shipping Agent | Freight Forwarders | Banks | Insurance | Regulators | Blockchain | Non-Blockchain (Traditional) | Time to Market |
|-----|--------------|----------|----------|---------|----------------|-------|----------|----------------|--------------------|-------|-----------|------------|------------|------------------------------|----------------|
| 16 | Faster settlement of disputes | ● | ● | ● | ● | ● | ● | ● | | ● | ● | ● | ✓ | ✓ | Blockchain>> Non-Blockchain |
| 17 | Risk management | ● | ● | ● | | | | | | | ● | ● | ✓ | | Blockchain>> Non-Blockchain |
| 18 | Provenance of goods | ● | ● | ● | ● | ● | ● | ● | ● | ● | ● | ● | ✓ | ✓ | Blockchain>> Non-Blockchain |
| 19 | Automatic reconciliations | ● | ● | ● | ● | ● | ● | ● | ● | ● | ● | ● | ✓ | ✓ | Blockchain>> Non-Blockchain |
| 20 | Document management | ● | ● | ● | ● | ● | ● | ● | ● | ● | ● | ● | ✓ | ✓ | Blockchain> Non-Blockchain |
| 21 | Built-in data recovery and backup | ● | ● | ● | ● | ● | ● | ● | ● | ● | ● | ● | ✓ | | Blockchain>> Non-Blockchain |
| 22 | Consensus | ● | ● | ● | | ● | ● | | | | | ● | ✓ | | Blockchain>> Non-Blockchain |
| 23 | Social scorecard | ● | ● | ● | ● | | | | | ● | ● | ● | ✓ | | Blockchain> Non-Blockchain |

Notes:

1. ● indicates that the item was identified as a requirement by the stakeholder.
2. In the three right-most columns, the requirements taken into consideration, independent of any technical considerations, are evaluated on three parameters: (i) can be implemented using blockchain technology; (ii) can be implemented using traditional technologies; and (iii) time to market, using either blockchain or traditional technologies. A ü indicates that blockchain or non-blockchain technology is capable to address the requirement.
3. "Blockchain>non-blockchain" denotes that using blockchain technology, the requirement can be implemented approximately 30% faster compared to non-blockchain (traditional) technology.
4. Non-blockchain>blockchain" denotes that using traditional technologies, the requirement can be implemented approximately 30% faster compared to blockchain technology.
5. "Blockchain>>non-blockchain" denotes that using blockchain technology, the requirement can be implemented approximately 50%–60% faster compared to non-blockchain (traditional) technology.

Source: Author's analysis.

network members. This leads to frictionless interactions among various participants and an overall improvement in efficiency.

## Data Privacy

Data privacy is a quintessential requirement for any international trade system. User information, confidential trade-related data, or any combination of information that can be used to identify any natural person must be protected. While traditional technologies can also offer data protection requirements in conjunction with industry-leading security products, blockchain offers such capability as a default functionality, with much faster development cycles.

## Domestic Trade Process Automation

If only the domestic trade process automation scope is taken into account, both traditional and blockchain technologies can address the requirements.

Business processes can be automated using both technologies, with reduced manual interventions. However, as blockchain offers distributed data networks and flexibility to integrate using multiple protocols with built-in channel mechanisms for security and privacy, the overall development effort required to fulfill all the requirements would be less using blockchain.

## Transaction Processing

Any network platform that independently cannot support transaction processing is an incomplete platform. An international trade network constitutes a transaction system that includes payment of fees, duties, fines, service charges, and settlements.

While the capabilities of transaction processing can be achieved using blockchain or traditional technologies, the underlying settlement mechanism, security protocols, and movement of ledger accounts can be achieved much more efficiently and using less development effort by using blockchain technology.

## Immutability

The trust and transparency of any system is dependent on the auditability of the system. Blockchain offers immutable records for any transaction at an unparalleled level. Any records added to the network cannot be removed or altered. This makes any blockchain solution tamperproof and reliable.

While using traditional technologies, audit records can be maintained, but data alterations are limited to access controls rather than immutability of the system.

## International Trade Process

While domestic trade process automation depends largely on consolidation and streamlining of domestic systems and processes, international trade process automation depends on data segregation, data standardization, data privacy, safeguarding of proprietary (or country-specific) business rules, and cross-border exchange of ownership of documents and payments.

These requirements can be addressed by using blockchain technology, as it offers capabilities to integrate cross-border systems for document exchange and transaction processing without compromising data privacy or security issues.

## Analytics and Cognitive Capabilities

A blockchain solution offers seamless integration with cognitive solutions, including risk management and predictive analytics. While traditional technologies can also provide such added functionalities, the quality and trustworthiness of the underlying data make the blockchain solution more reliable.

## Efficient Fund Management

A distributed ledger facilitates the movement of ownership of assets and documents. By using blockchain technology, fund movement between network members can be settled in near real time, thereby improving the overall efficiency of working capital management. This allows network members to maintain a single account and settle payment obligations in near real time rather than maintaining pre-funded accounts with multiple parties.

## Faster Settlement of Disputes

Blockchain maintains records of every transaction or event on a shared ledger platform. These data are only added to the ledger once consensus is achieved among the network participants. As disputes arise in case of conflict of information or data mismatch at the time of declaration or payments, such disputes can be completely eradicated using blockchain.

## Risk Management

A permissioned blockchain network is built on both an operational and a legal construct. The legal construct mandates each participant to own responsibility of these data uploaded to the distributed ledger. This makes the blockchain a network of accountability for the shared data.

Each permissioned network member is responsible and accountable for these data or documents loaded to the ledger, thereby reducing the risk of fraud or data tampering. This leads to improved risk management from the perspective of managing events, actions, and transactions on the network.

## Provenance of Goods

For a truly international trade platform, the provenance of goods, established at the source, is of paramount importance. This feature vastly improves transparency and trust in the system. Importers can trust the origin of the product, customs officials can correctly determine the pricing of products, and regulators can verify the associated certifications from the source systems.

## Automatic Reconciliations

As documents and data are added to the blockchain ledger only after the consensus protocols are successfully executed, the ledger data is referred to as "gold standard data."

This greatly reduces the need to run reconciliation programs to match data across multiple systems. Now these data on the blockchain network can be considered as the correct data, and any peripheral application that requires replication can automatically synchronize with the gold standard data.

### Built-in Data Recovery and Backup

A distributed system like blockchain offers default functionalities for data backup at multiple nodes and automatic backup protocols. Data always remain in sync among replicated ledgers and with high availability and an inbuilt fail-safe mechanism, i.e., data are never lost.

### Network Process Standardization

A blockchain design and development exercise starts with data standardization. This is a key and fundamental element for building blockchain networks, leading to data harmonization across multiple organizations, within or outside of national boundaries. While traditional technologies can also address this requirement, blockchain technology makes this effort seamless and more efficient.

# B.2  Blockchain versus Traditional Technologies

In the context of building a cross-border trade network, Table 3 provides a comparison between traditional and blockchain technologies. The key difference in building a truly international trade system, using either a traditional approach or a blockchain-based approach, is in the underlying design consideration.

A blockchain-based approach is focused on building a secure, shared, and distributed ecosystem from the ground up and from day one. On the other hand, a traditional approach is more aligned with a scale-up model where the base applications are built for a single country and then extended to integrate with cross-border systems.

A blockchain application offers seamless domestic and cross-border integration capabilities with multiple applications, allowing user data to be secure, encrypted, and immutable to preserve trust and transparency.

Furthermore, a blockchain-based system offers higher levels of availability and scalability. It allows faster onboarding of network members with different roles without affecting the overall functional design or performance.

A key concern for Maldives is the availability of disaster recovery sites in the country. While a single database with an offsite disaster recovery location on a different island is better than hosting the disaster recovery site at the same location of the primary database (as per current practice), a distributed ledger technology with a cloud-hosted recovery mechanism offers a more robust data backup and recovery solution.

**Table 3: Comparison between Traditional and Blockchain Technologies**

| No. | Area | Traditional Technologies | Blockchain |
|-----|------|--------------------------|------------|
| 1 | Central database | Single point of failure | Distributed ledger; no single point of failure |
| 2 | Trust and transparency | Access control via traditional roles and responsibilities | Immutable ledger, permissioned network; non-trusted parties can also be onboarded |
| 3 | Data control | Central control; prone to tampering | Distributed control; participants control their own data without compromising on privacy; no data tampering possible |
| 4 | Middleware | Single layer to be managed by one central party or organization | Middle layer can be created, managed, and controlled by network, giving participants more control over their data and business rules |
| 5 | Off-chain database | Single common database for document data and document management | Data and documents are encrypted and loaded to the shared ledger, while documents are either maintained at source systems or in off-chain databases; this considerably improves the query process and overall performance of the system |
| 6 | Load distribution and performance | Single database and common application layer | Load distribution through redundant and distributed nodes with multiple middleware layers |
| 7 | Data security and privacy | Data secured through traditional technology protocols | Data security and privacy achieved through inbuilt encryption mechanisms and distributed ledger protocols, which provides a tamperproof and immutable system |
| 8 | High availability | Availability restricted by single databases and common application layers | High availability through distributed network protocols and redundant nodes |
| 9 | Scalability | Scalability is restricted by infrastructure provisioning | Highly scalable design and future-proof from the perspective of adding new organizations or business rules to the network |

Note: Middleware is a software that acts as a bridge between an operating system or database and applications, especially on a network.

Source: Author's analysis.

# B.3  Description of Blockchain Use Cases

The following use cases were identified as part of "Study on Using Blockchain Technology for Paperless Trade Facilitation in Maldives, under South Asia Subregional Economic Cooperation National Single Window Project," that can address the current challenges and satisfy key requirements for a truly international trade system. Six key use cases identified during the study are listed in Table 4.

## Table 4: Blockchain Use Cases

| No. | Blockchain Use Case | Description |
|---|---|---|
| 1 | Single platform for trade document exchange | Trade participants (such as the consignee, customs broker, shipping agent, freight forwarder, vessel agent, MACL, MCS, and MPL have to be engaged through multiple manual and digital channels to exchange trade-related documents. Since most of these documents are exchanged manually, the inherent redundancies, repetitions, and operational inefficiencies in the overall process are considerably high.<br><br>Blockchain can offer a single trade document exchange platform that allows for single submission of documents (including bills of lading, cargo manifests, second carrier bills of lading, invoices, purchase orders, packing lists, air waybills, and inward lists), by the relevant network participant. This considerably reduces the time and effort taken to share these documents in a point-to-point exchange mechanism.<br><br>Using blockchain, documents are digitized and a single version of each document is available only to the relevant trade participant. |
| 2 | Single platform for import permit processing and issue | Currently, the trader has to engage with multiple government agencies, including the MED, MCS, MFMRA, MFDA, and MOD, to apply for relevant permits. While some of the agencies offer partial digital solution for download of application forms and upload of documents, an end-to-end, transparent application process for the issue of digital permits is still not available.<br><br>Also, the permit issue process requires collaboration between multiple government agencies, which is currently done manually. The applicant has to visit multiple agencies, one by one, to secure approvals.<br><br>The applicant also needs to pay each agency separately for the processing charges. A single blockchain-based platform for permit application processing and issue would greatly improve the customer experience and reduce turnaround time. Such solution would also lead to improved process level transparency, a seamless workflow among different government agencies, and automated payments management. |
| 3 | Single platform for CBRAs: clearance of goods | The role of CBRAs in the international trade process is of paramount importance, as CBRAs have a range of responsibilities, including enforcing compliance standards of export products and border control for goods moving into the country. From a trade perspective, CBRAs play the role of issuing permits and certificates, and examining goods at ports and airports.<br><br>A blockchain-based solution would allow for a workflow-based mechanism where for each delivery order the CBRAs are mapped. An advance notification to the concerned CBRAs will be sent for goods inspection. At the point of physical inspection, the entries can be inputted directly to the distributed ledger that can be viewed by other CBRAs, MPL, and MACL. This would also remove the current process of physical stamping of gate pass (release goods note at MPL). |
| 4 | Single payments account | While online payment options are available in Maldives, most of the trade-related payments are still made in either check or cash. This is primarily due to the high transaction costs involved in online payments. As the current trade economy is unable to take advantage of real-time liquidity provisions, most traders, customs brokers, shipping agents, and freight forwarders are left with the choice of maintaining a pre-funded account with MACL, MCS, and MPL. |

*continued on next page*

*Table 4 continued*

| No. | Blockchain Use Case | Description |
|---|---|---|
| | | This leads to considerable liquidity lock-in and inefficient use of working capital. A blockchain-based solution would allow for near real-time settlement of accounts (ledgers) between the participating bank, the service provider (government agency and private sector) and the customer (trader, shipping agent, customs broker). With the use of a single blockchain platform, customers do not have to maintain separate accounts with multiple service providers but one account with the bank on the blockchain network. |
| 5 | International trade platform | In the current international trade process, document and data exchange happens between traders, trade facilitators, and government agencies. Noncompliance is penalized with fines and there are delays in the movement of goods. |
| | | The lead time for the arrival of goods from Sri Lanka to Maldives is about 1 hour by air and about 36 hours by sea. This leads to serious strain on existing trade processes to complete the required documentation and plan for the arrival of cargo. Because of limited land mass (offloading or inspection area, transport loading), there is an urgency to move containers out of the landing zones. |
| | | Due to the time constraints and manual documentation procedures, the risk of fraud also increases. There have been incidents where traders have declared lower prices for incoming goods, leading to lower duty payments. |
| | | An international blockchain network for trade would significantly improve the following: |
| | | (i) Risk management would be improved since invoices, purchase orders, and other price determination documents (if any) could be traced back to the source. This would reduce the chances of document fraud (incorrect price declaration and determination) and improve confidence in the system. |
| | | (ii) Cross-border collaboration between government agencies and port and airport operators. |
| | | (iii) Exchange of such documents as export or import declarations and manifests would improve trust and transparency in the system. Trade documents would be automatically generated or uploaded on the blockchain network. There would be considerable reduction in any manual intervention and any changes would be captured on the blockchain network. Port or air cargo operators would receive the single version of trade documents like bills of lading, second carrier bills of lading, manifests, packing lists, the moment it is uploaded or generated on the blockchain network. This would also eliminate the need to share multiple copies of same document with different organizations. |
| | | (iv) The visibility of banks into the trade process would be improved through improved inputs for risk assessment of customers during the letter of credit issue process. |
| 6 | Single platform for export permits and certifications | Blockchain use case 2 (single platform for import permit processing and issue) describes the benefits of having a blockchain-based import permit processing and issue process. This process can be logically extended to include the processing and issue of export permits and certifications. |
| | | Catch certificates are issued by the MFMRA and health certificates are issued by MFDA. The applicant has to pay for these certificates manually by visiting the offices of these entities multiple times. The application process is also manual in nature without any workflow automation for inter- or intra-organization collaboration. |

*continued on next page*

*Table 4 continued*

| No. | Blockchain Use Case | Description |
|---|---|---|
| | | A blockchain-based system would consolidate the process for export permits and certification onto a single platform where the applicant only needs to apply and pay for the fee once. |

CBRA = cross-border regulatory agency; MACL = Maldives Airports Company Limited; MCS = Maldives Customs Service; MED = Ministry of Economic Development; MFDA = Maldives Food and Drug Authority; MFMRA = Ministry of Fisheries, Marine Resources and Agriculture; MOD = Ministry of Defence; MPL = Maldives Ports Limited.

Source: Author's analysis.

# B.4 Blockchain as National Single Window Backbone

Figure 3 depicts the comparison between a traditional national single window (NSW) architecture and an NSW architecture with blockchain as the backbone. This depiction also establishes the applicability of an international blockchain network, in the context of current NSW architecture.

Typical NSW architectures are designed and built for domestic trade ecosystems. An NSW architecture offers a single portal to access different services, and it follows a traditional three-layer structure with web servers, applications, and a central database.

On the other hand, blockchain is designed as a distributed network solution, consisting of domestic and cross-border participants. It offers a single digital workflow across domestic organizations, and network workflows across cross-border agencies.

A blockchain network can seamlessly integrate with existing NSW architectures across multiple countries and regions, providing a common platform to share data and documents, and to transact through inbuilt ledger movements.

Some of the key features of a blockchain network application in the context of being a backbone for multiple NSW implementations across different countries are described in the following paragraphs.

**It does not disrupt the national single window architecture.** A blockchain solution does not disrupt the NSW architecture or any other domestic systems or processes that may be required to integrate with the backbone. A blockchain solution provides API endpoints for easy integration and also supports file-based transfers.

**It offers superior performance and data security.** Blockchain, with redundant node architecture, offers high fault tolerance, with superior system durability and availability.

**It is future proof.** Blockchain is a proven technology for the future and is best suited to build domestic and international business networks. It offers improved levels of network automation, immutability, and scalability features.

**It improves overall trust in the system.** Blockchain as an NSW backbone improves overall trust and transparency in the international trade ecosystem. It allows future participants to join the network seamlessly, without any friction arising from cross-border exchange of data. A single digital workflow across national boundaries allows for greater cooperation and efficiencies.

## Figure 3: Blockchain Layer as National Single Window Backbone

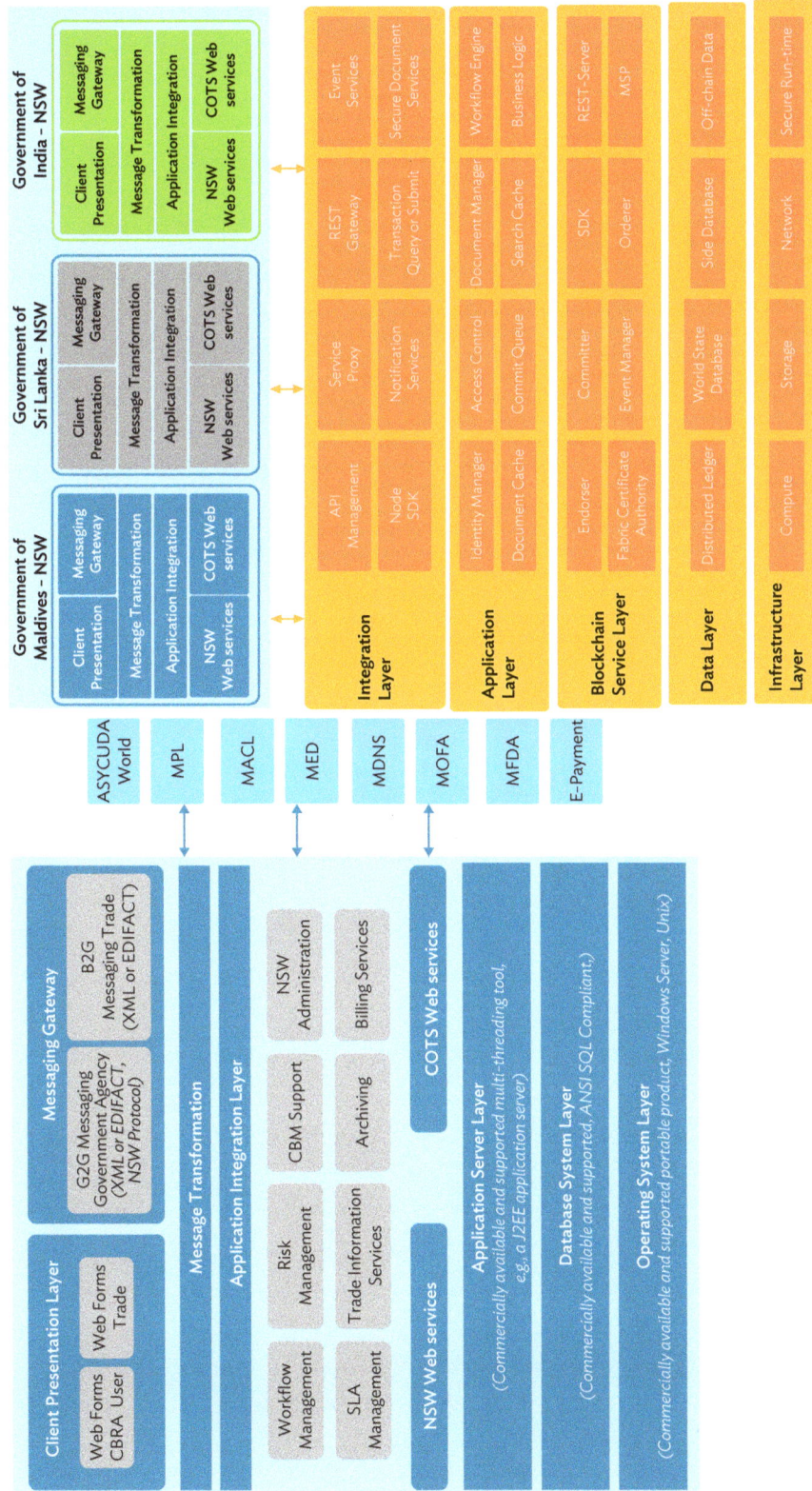

ANSI = American National Standards Institute, API = application programming interface, ASYCUDA = Automated Systems for Customs Data, B2G = business-to-government, CBM = coordinated border management, CBRA = cross-border regulatory agency, COTS = commercial off-the-shelf, G2G = government-to-government, J2EE = Java 2 platform enterprise edition, MACL = Maldives Airports Company Limited, MED = Ministry of Economic Development, MFDA = Maldives Food and Drug Authority, MOFA = Ministry of Foreign Affairs, MPL = Maldives Ports Limited, MSP = membership service provider, NSW = national single window, REST = representational state transfer, SDK = software development kit, SLA = service-level agreement.

Source: Author's analysis.

**Data encryption and security.** Data added to the blockchain ledger are tamperproof and cannot be removed or altered in any way. This makes the ledger immutable and trustworthy. In many blockchain implementations, the ledger data is considered a "golden source of truth" and enables faster settlement of disputes and real-time visibility and transparency of operations.

# Road Map for Blockchain Adoption in Maldives

## C.1 Adoption Road Map

The road map for blockchain adoption starts with a successful pilot. The use case for such a pilot must be selected carefully, and with consensus among key stakeholders. A successful blockchain pilot must be able to demonstrate the technological capabilities of a distributed system. At the same time, it must satisfy key requirements related to operational performance improvement and the strategic value of long-term investment in cutting-edge technology platforms. Figure 4 depicts a four-step road map adoption model.

### Figure 4: Blockchain Adoption Road Map

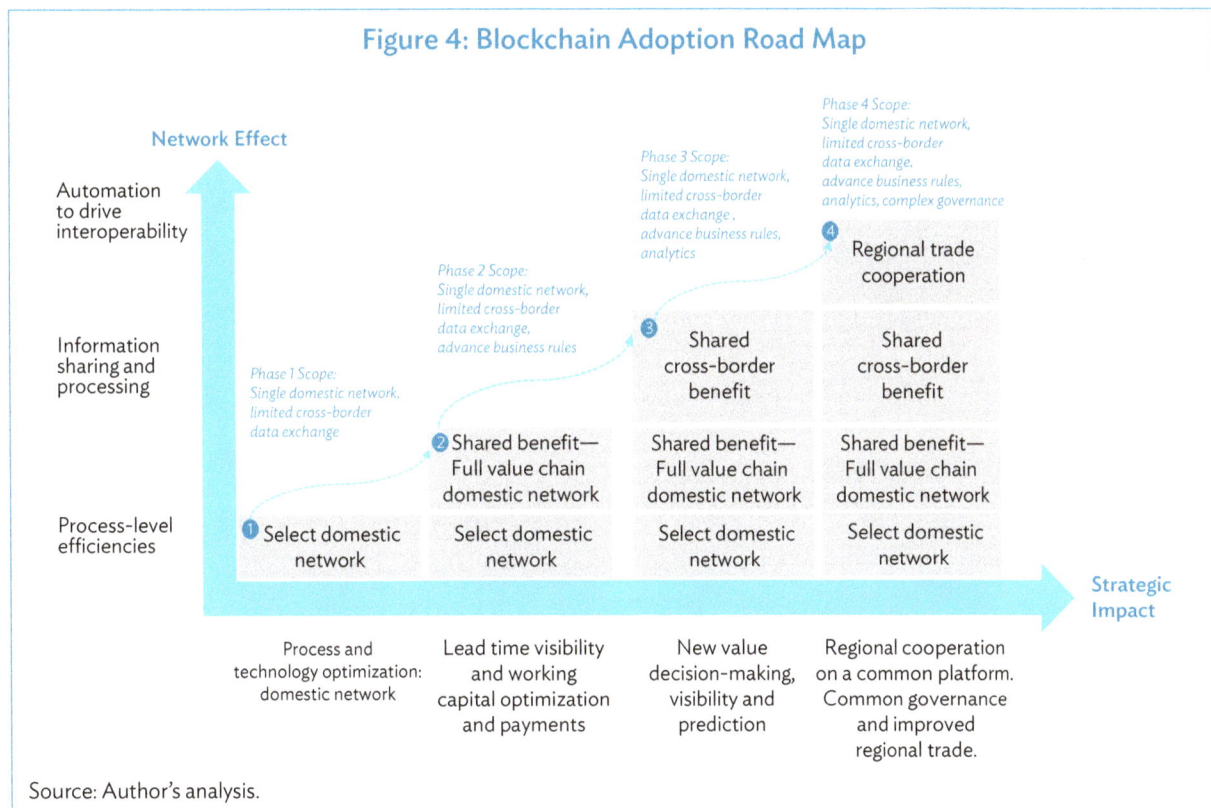

**Network Effect** (vertical axis)

Automation to drive interoperability

Information sharing and processing

Process-level efficiencies

**Strategic Impact** (horizontal axis)

*Phase 1 Scope:*
*Single domestic network, limited cross-border data exchange*

*Phase 2 Scope:*
*Single domestic network, limited cross-border data exchange, advance business rules*

*Phase 3 Scope:*
*Single domestic network, limited cross-border data exchange, advance business rules, analytics*

*Phase 4 Scope:*
*Single domestic network, limited cross-border data exchange, advance business rules, analytics, complex governance*

| | | | ④ Regional trade cooperation |
| | | ③ Shared cross-border benefit | Shared cross-border benefit |
| | ② Shared benefit— Full value chain domestic network | Shared benefit— Full value chain domestic network | Shared benefit— Full value chain domestic network |
| ① Select domestic network | Select domestic network | Select domestic network | Select domestic network |
| Process and technology optimization: domestic network | Lead time visibility and working capital optimization and payments | New value decision-making, visibility and prediction | Regional cooperation on a common platform. Common governance and improved regional trade. |

Source: Author's analysis.

Phase 1 involves designing, developing, and testing of a pilot blockchain application for a select use case. Design principles for the pilot must take into consideration the requirements for a fully operational platform in the future. These include node scalability, smart contracts to facilitate a modular functional structure, easy integration through representational state transfer APIs, and an easy onboarding mechanism.

A well-designed pilot application would allow for rapid scaling during phase 2, when the distributed network is expanded to include all key participants from the domestic trade ecosystem.

During phases 3 and 4, the blockchain platform is extended to cross-border participation. The success of these phases depends on the shared benefits from the ecosystem. Different members of the network should be able to derive value from the shared network. For example, a customs agency must be able to benefit from the cross-border sharing of information from international counterparts. The immutability and security of the platform should allow various border control agencies to provide early warning notices to concerned domestic and international parties. An importer should have absolute transparency in goods declaration, duties, inspection, clearance, and payment of dues.

A regional blockchain network creates a single platform for governments and private agencies, where members can securely transact and share data.

# C.2 Blockchain Pilot

Blockchain use cases are described in detail in section B.3.

Each use case is evaluated on two parameters: (i) feasibility of development, and (ii) strategic impact.

Figure 5 depicts the mapping of each use case based on strategic impact and feasibility (across two axes).

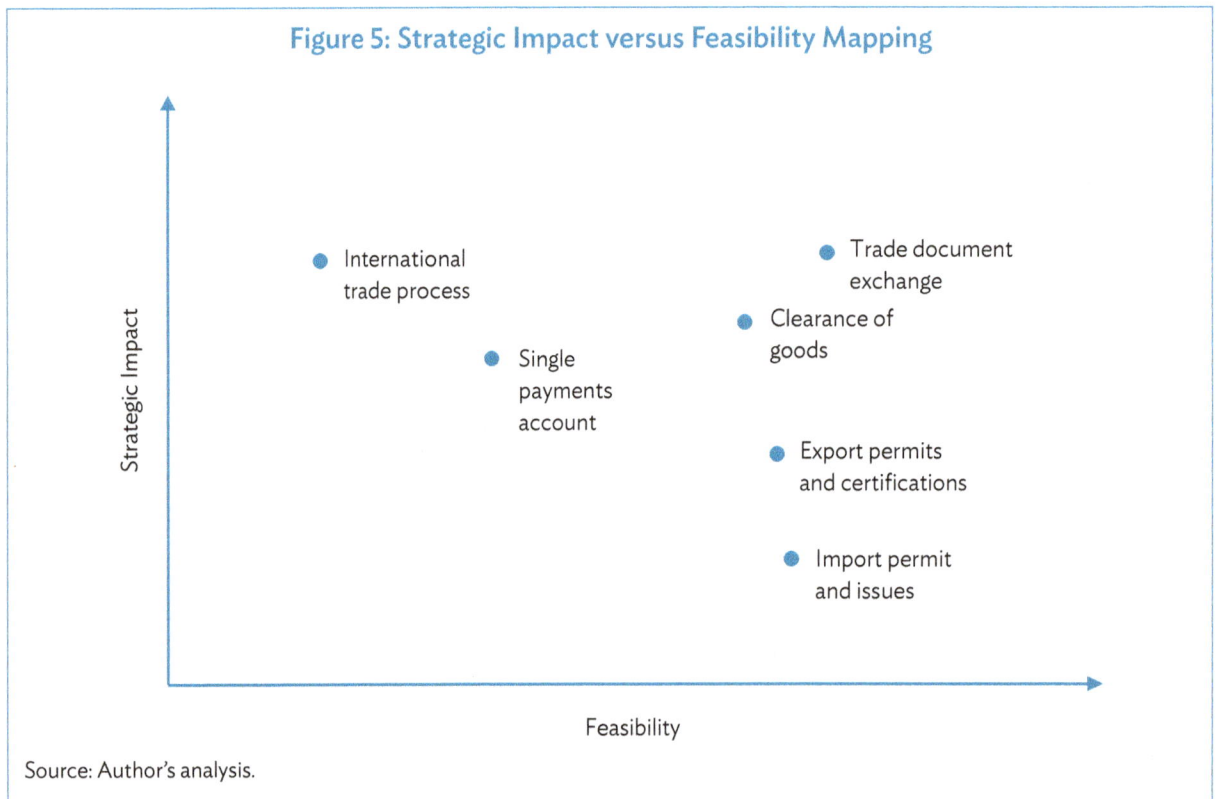

**Figure 5: Strategic Impact versus Feasibility Mapping**

Source: Author's analysis.

The use case for trade document exchange was selected based on the comparison between strategic impact and feasibility mapping (Figure 5) and consensus among key stakeholders.

## Network Members for Blockchain Pilot

The following organizations will participate in the blockchain pilot:

(i)    customs broker (one customs broker in Maldives);
(ii)   shipping agent (one shipping agent in Maldives);
(iii)  freight forwarder (one freight forwarder, which could be the same as the shipping agent);
(iv)   MPL;
(v)    MCS; and
(vi)   one cross-border customs agency to handle the customs application for exchange of documents (limited to cargo manifest document exchange).

## Process Map

The scope of the blockchain pilot is shown in Figure 6. In this process, all the participants of the blockchain trade network are able to digitally share data and documents. These data are immutably stored on the blockchain ledger and documents are cryptographically indexed in the ledger. To enable faster read or write capabilities, documents are stored on an off-chain database component and cryptographically linked to the ledger.

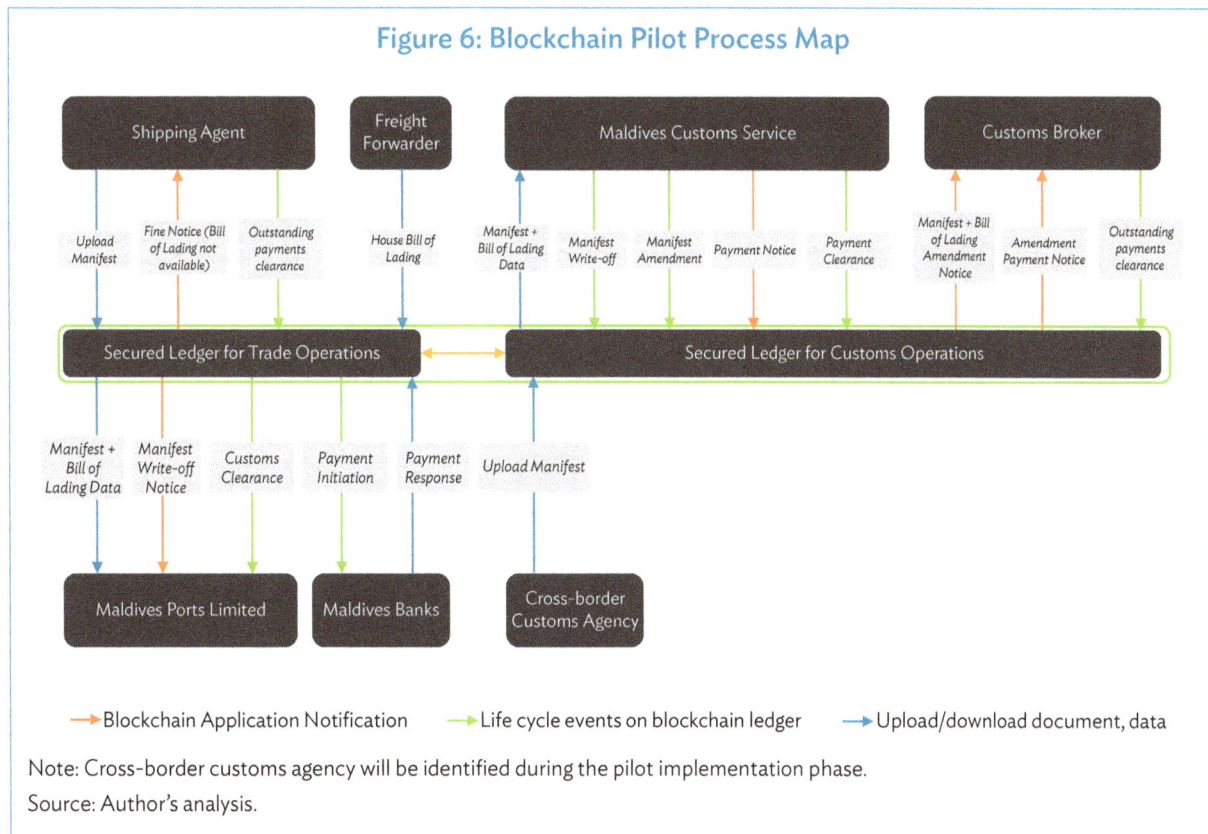

### Figure 6: Blockchain Pilot Process Map

Note: Cross-border customs agency will be identified during the pilot implementation phase.

Source: Author's analysis.

A cross-border customs agency would also be able to participate in this network, allowing seamless coordination between Maldives and the overseas customs agency to have a single view into any trade document (e.g., a cargo manifest).

The pilot scope also includes API-based integration between the blockchain network and domestic banks in Maldives. This would allow the participants to trigger online payment instructions from the blockchain user interface, and to receive payment confirmation notices.

## Timelines

The blockchain pilot will be completed over a period of 16 weeks, including user acceptance testing and integration testing. The pilot will be hosted and supported for another 4 weeks to determine realized benefits and feedback.

## Technical Architecture

The technical architecture for the blockchain pilot is depicted in Figure 7. It consists of the following components:

(i) **Access layer.** User access to the blockchain solution will be managed through this layer. In the background, access control lists will be maintained with user profiles and role descriptions.

(ii) **User interface layer.** This layer will allow users to enter trade-related data, upload documents, and view notifications. It will be developed using technologies like AngularJS and Express JS.

(iii) **Application layer.** This layer will serve as the backbone for the user interface by providing the services for workflow management, documents management, and events management. It will be built using technologies like Node.js.

(iv) **Blockchain service layer.** This layer will support the services of blockchain and the management of smart contracts, consensus, and endorsements. This layer will also support identity management for blockchain access. This layer will be built using Hyperledger Fabric 2.0.

(v) **Data layer.** This layer hosts the distributed ledger to store encrypted blocks, along with the latest transactions data. This layer also maintains off-chain databases to manage user credentials and business metadata information. This layer will be developed using Hyperledger Fabric. 2.0 and IBM Cloudant.

(vi) **Integration layer.** This layer is hosted in a Node.js runtime environment for managing all the services, exposed by a blockchain application layer. This layer also acts as a service proxy for managing integration with the system of records. This layer will be developed using Node.js.

# C.3 Pilot Value Proposition and Key Outcomes

The success of any blockchain network depends on the benefits sharing model; each participant should be able to benefit from the shared network. For the proposed blockchain network, the benefits sharing model is depicted in Table 5.

Participants in the blockchain trade network would be able to benefit directly through, for example, a reduction in effort, delays, and errors with manual payments. Indirect benefits include timely availability of information related to amendments due to the second carrier bill of lading.

## Figure 7: Blockchain Pilot Application Architecture

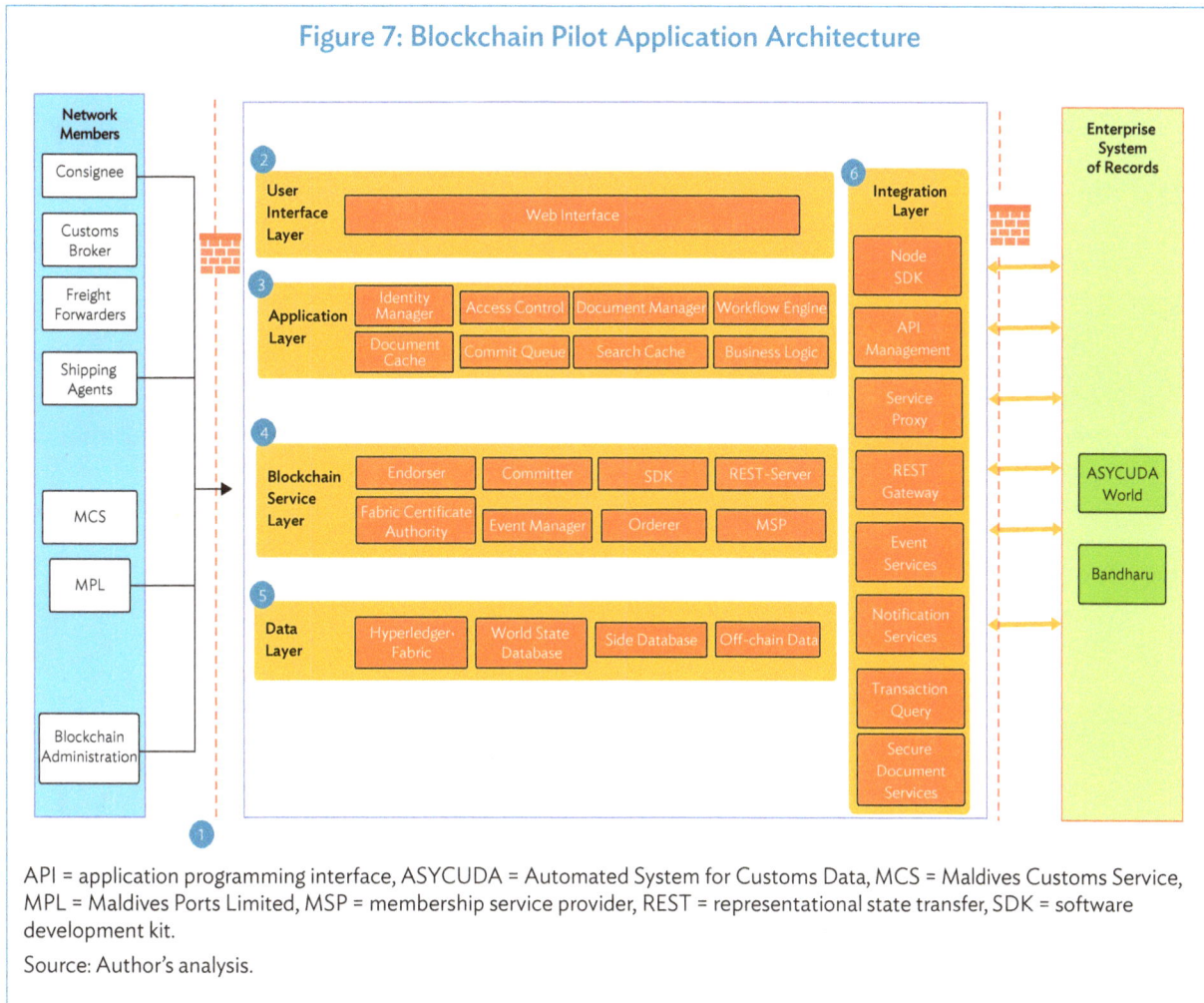

**Network Members**
- Consignee
- Customs Broker
- Freight Forwarders
- Shipping Agents
- MCS
- MPL
- Blockchain Administration

**User Interface Layer** (2)
- Web Interface

**Application Layer** (3)
- Identity Manager
- Access Control
- Document Manager
- Workflow Engine
- Document Cache
- Commit Queue
- Search Cache
- Business Logic

**Blockchain Service Layer** (4)
- Endorser
- Committer
- SDK
- REST-Server
- Fabric Certificate Authority
- Event Manager
- Orderer
- MSP

**Data Layer** (5)
- Hyperledger Fabric
- World State Database
- Side Database
- Off-chain Data

**Integration Layer** (6)
- Node SDK
- API Management
- Service Proxy
- REST Gateway
- Event Services
- Notification Services
- Transaction Query
- Secure Document Services

**Enterprise System of Records**
- ASYCUDA World
- Bandharu

API = application programming interface, ASYCUDA = Automated System for Customs Data, MCS = Maldives Customs Service, MPL = Maldives Ports Limited, MSP = membership service provider, REST = representational state transfer, SDK = software development kit.

Source: Author's analysis.

## Table 5: Blockchain Benefits Sharing Table

| Value Proposition | Customs Broker | Shipping Agent | Freight Forwarder | Maldives Customs Service | Maldives Ports Limited | Overseas Customs Agency |
|---|---|---|---|---|---|---|
| Single view of bill of lading | ✓ | ✓ | ✓ | ✓ | ✓ | |
| Single view of cargo manifest | ✓ | | ✓ | ✓ | ✓ | ✓ |
| Digital documents sharing | ✓ | ✓ | ✓ | ✓ | ✓ | ✓ |
| Near real-time visibility in trade life cycle | ✓ | ✓ | ✓ | ✓ | ✓ | |
| Automated events notifications | ✓ | ✓ | ✓ | ✓ | ✓ | |
| Real-time payment notifications | ✓ | ✓ | ✓ | ✓ | ✓ | |
| Real-time integration with bank payment portal | ✓ | ✓ | ✓ | ✓ | ✓ | |

Source: Author's analysis.

The following are the key outcomes that will be offered, as part of the pilot, to the entire trade ecosystem:

(i)   Automated workflows to manage the document exchange process.

(ii)  Digital transfer of document ownership, e.g., ownership transfer of a delivery order from a shipping agent to a consignee.

(iii) Automated data validations using blockchain smart contracts. Any data entered on the blockchain user interface will be validated using pre-built rules, like matching existing bill of lading data with second bill of lading data.

(iv)  Interactive user interface for data entry and document upload. Network members would be able to use user interface screens to enter trade-related data or upload documents. For example, shipping agents should be able enter and update data related to ETA and payment confirmation.

(v)   Reduced turnaround time and improved network service-level agreements. Network service-level agreements would be implemented for document exchange, payments, and amendments processes, leading to improved operational efficiencies.

(vi)  Improved trust in the network. As blockchain offers an immutable ledger and permissioned network member participation, the trust in the network will improve.

(vii) Reduction in cost. The blockchain pilot will improve operational efficiencies, with reductions in (a) redundancy and repetition in the document exchange process, and (b) fixed costs associated with data entry and data reconciliation.

## Figure A.1: Customs Registration

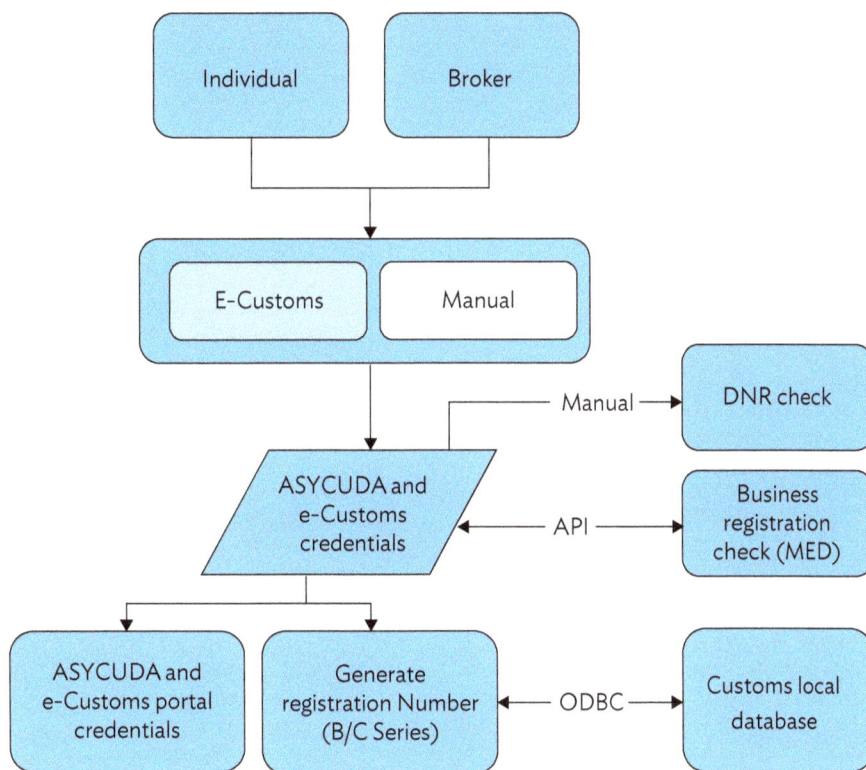

API = application programming interface, ASYCUDA = Automated System for Customs Data, B/C= B number for broker and C number for consignee, DNR = Department of National Registration, MED = Ministry of Economic Development, ODBC = open database connectivity.

Note: Two series numbers are generated at the time of registration. "B" series numbers are assigned to Customs Brokers and "C" series numbers are assigned to Consignee.

Source: Author's analysis.

## Figure A.2: Business Registration

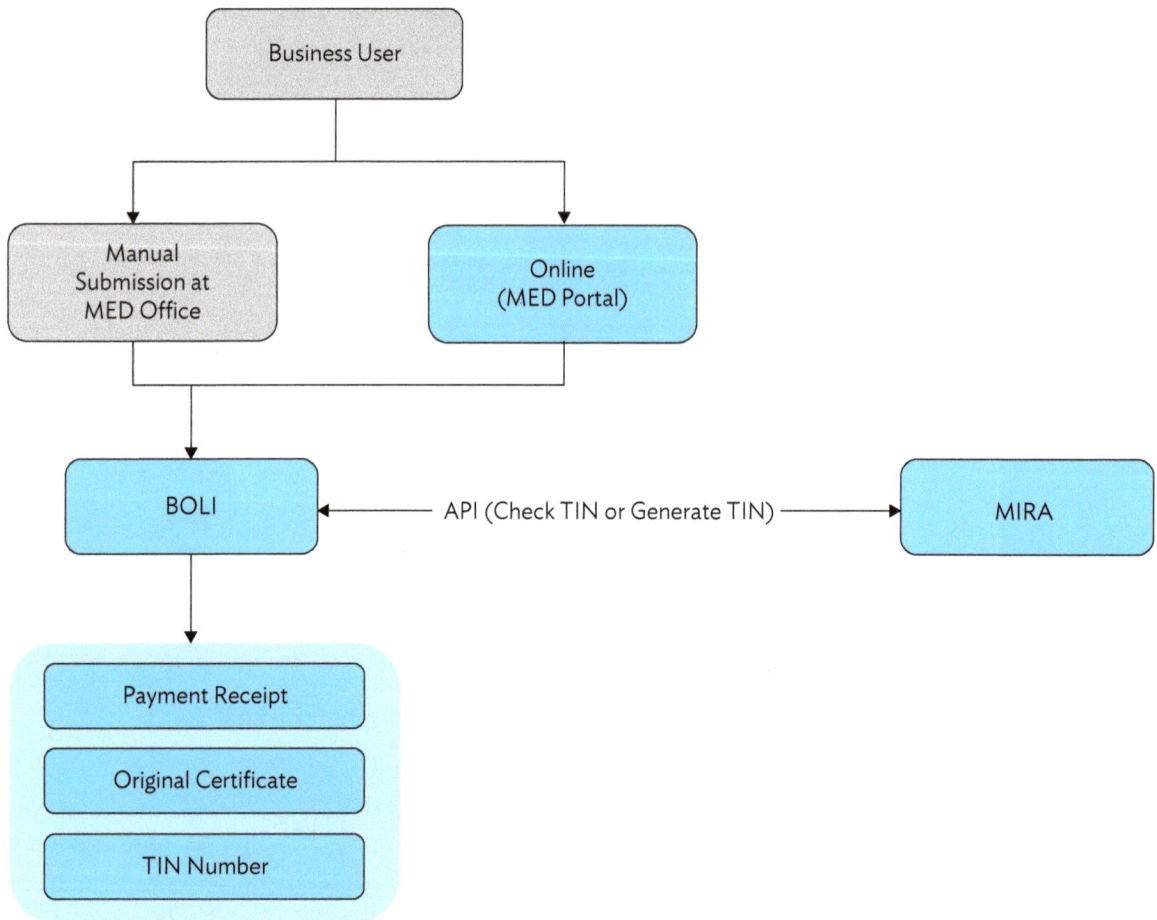

API = application programming interface, MED = Ministry of Economic Development, MIRA = Maldives Inland Revenue Agency, TIN = taxpayer identification number.

Note: BOLI is the online MED portal.

Source: Author's analysis.

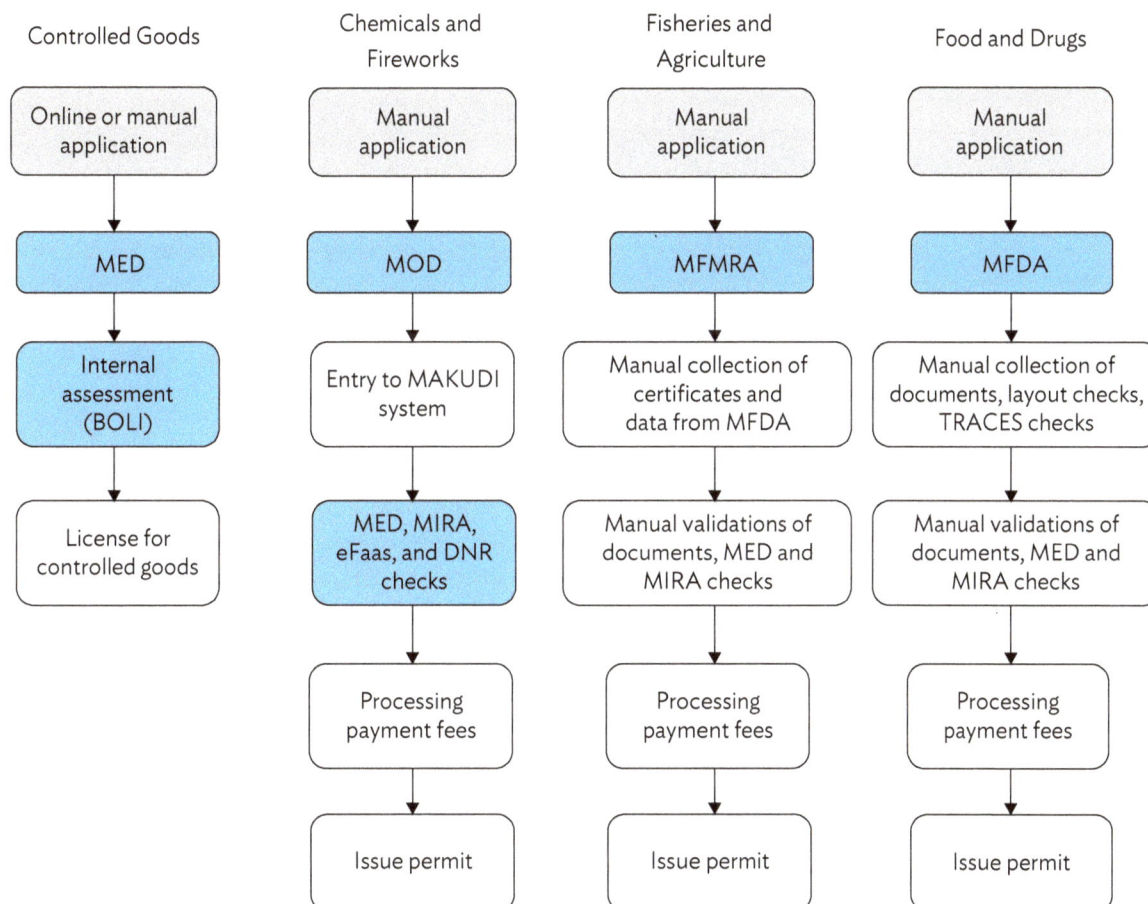

Figure A.3: Permits Process of Cross-Border Regulatory Agencies

DNR = Department of National Registration; MED = Ministry of Economic Development; MFDA = Maldives Food and Drug Authority; MFMRA = Ministry of Fisheries, Marine Resources and Agriculture; MIRA= Maldives Inland Revenue Authority; MOD = Ministry of Defence; TRACES = trade control and expert system.

Notes:
1. BOLI is the online MED portal.
2. MAKUDI refers to the IT system of MOD.

Source: Author's analysis.

## Figure A.4: Customs Declaration

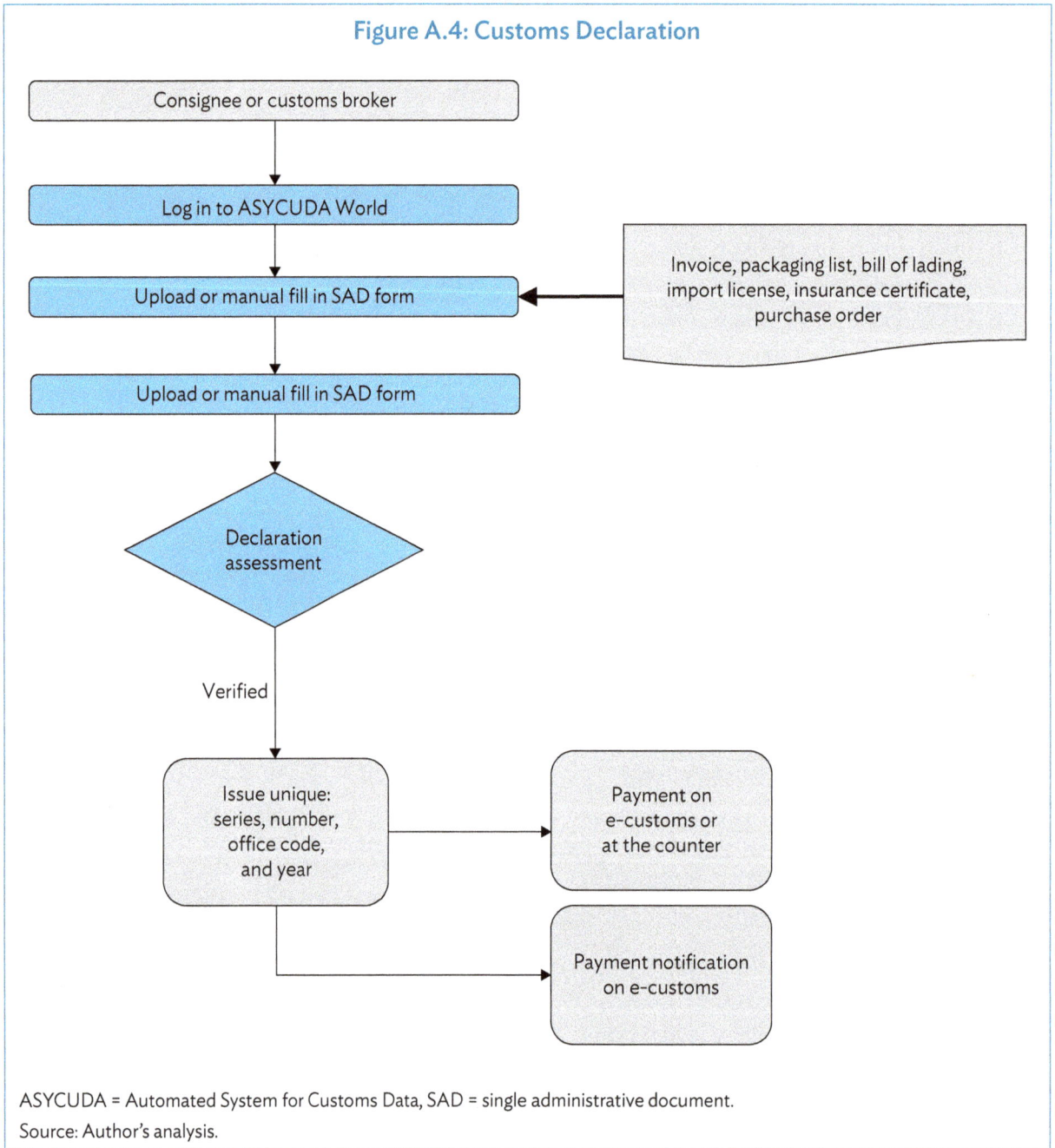

```
┌─────────────────────────────────┐
│   Consignee or customs broker   │
└─────────────────────────────────┘
                 │
                 ▼
┌─────────────────────────────────┐
│      Log in to ASYCUDA World    │
└─────────────────────────────────┘
                 │                          ┌──────────────────────────────────┐
                 ▼                          │ Invoice, packaging list, bill of │
┌─────────────────────────────────┐        │ lading, import license, insurance│
│   Upload or manual fill in SAD  │ ◄────── │ certificate, purchase order      │
│              form               │        └──────────────────────────────────┘
└─────────────────────────────────┘
                 │
                 ▼
┌─────────────────────────────────┐
│   Upload or manual fill in SAD  │
│              form               │
└─────────────────────────────────┘
                 │
                 ▼
          ╱─────────────╲
         ╱  Declaration  ╲
         ╲  assessment    ╱
          ╲─────────────╱
                 │
              Verified
                 │
                 ▼
┌──────────────────────┐         ┌──────────────────────┐
│   Issue unique:      │         │   Payment on         │
│   series, number,    │ ──────► │   e-customs or       │
│   office code,       │         │   at the counter     │
│   and year           │         └──────────────────────┘
└──────────────────────┘
                 │               ┌──────────────────────┐
                 └─────────────► │ Payment notification │
                                 │   on e-customs       │
                                 └──────────────────────┘
```

ASYCUDA = Automated System for Customs Data, SAD = single administrative document.
Source: Author's analysis.

## Figure A.5: Cargo Manifest

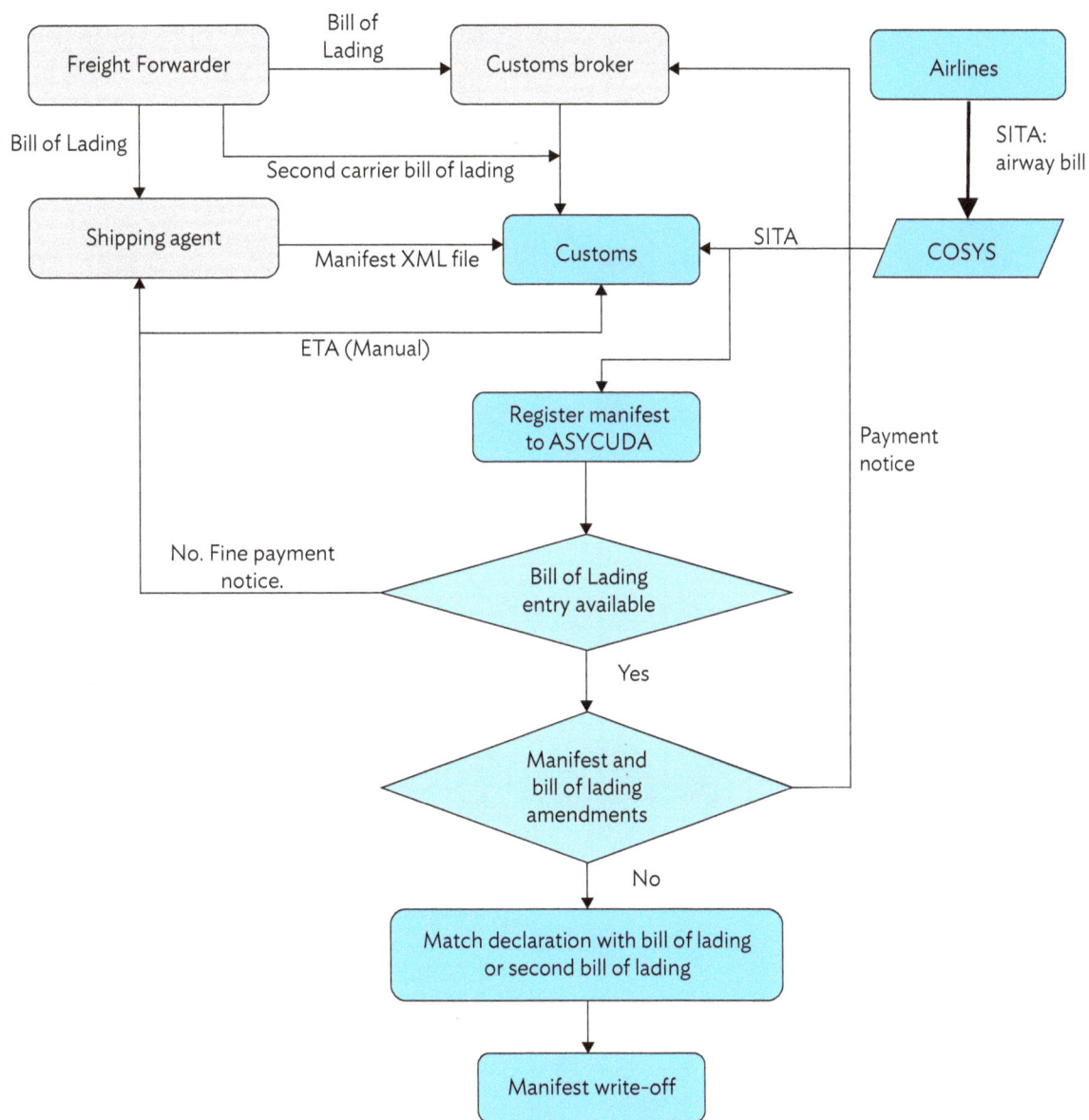

Freight Forwarder —Bill of Lading→ Customs broker ← Airlines

Airlines —SITA: airway bill→ COSYS

Freight Forwarder —Bill of Lading→ Shipping agent

Freight Forwarder —Second carrier bill of lading→ Customs

Shipping agent —Manifest XML file→ Customs

COSYS —SITA→ Customs

ETA (Manual)

Customs → Register manifest to ASYCUDA

Register manifest to ASYCUDA → Bill of Lading entry available

No. Fine payment notice.

Bill of Lading entry available —Yes→ Manifest and bill of lading amendments

Payment notice

Manifest and bill of lading amendments —No→ Match declaration with bill of lading or second bill of lading

Match declaration with bill of lading or second bill of lading → Manifest write-off

ASYCUDA = Automated System for Customs Data, COSYS = cargo operations system, ETA = estimated time of arrival, SITA = Société Internationale de Télécommunications Aéronautiques.
Source: Author's analysis.

## Figure A.6: Customs Clearance

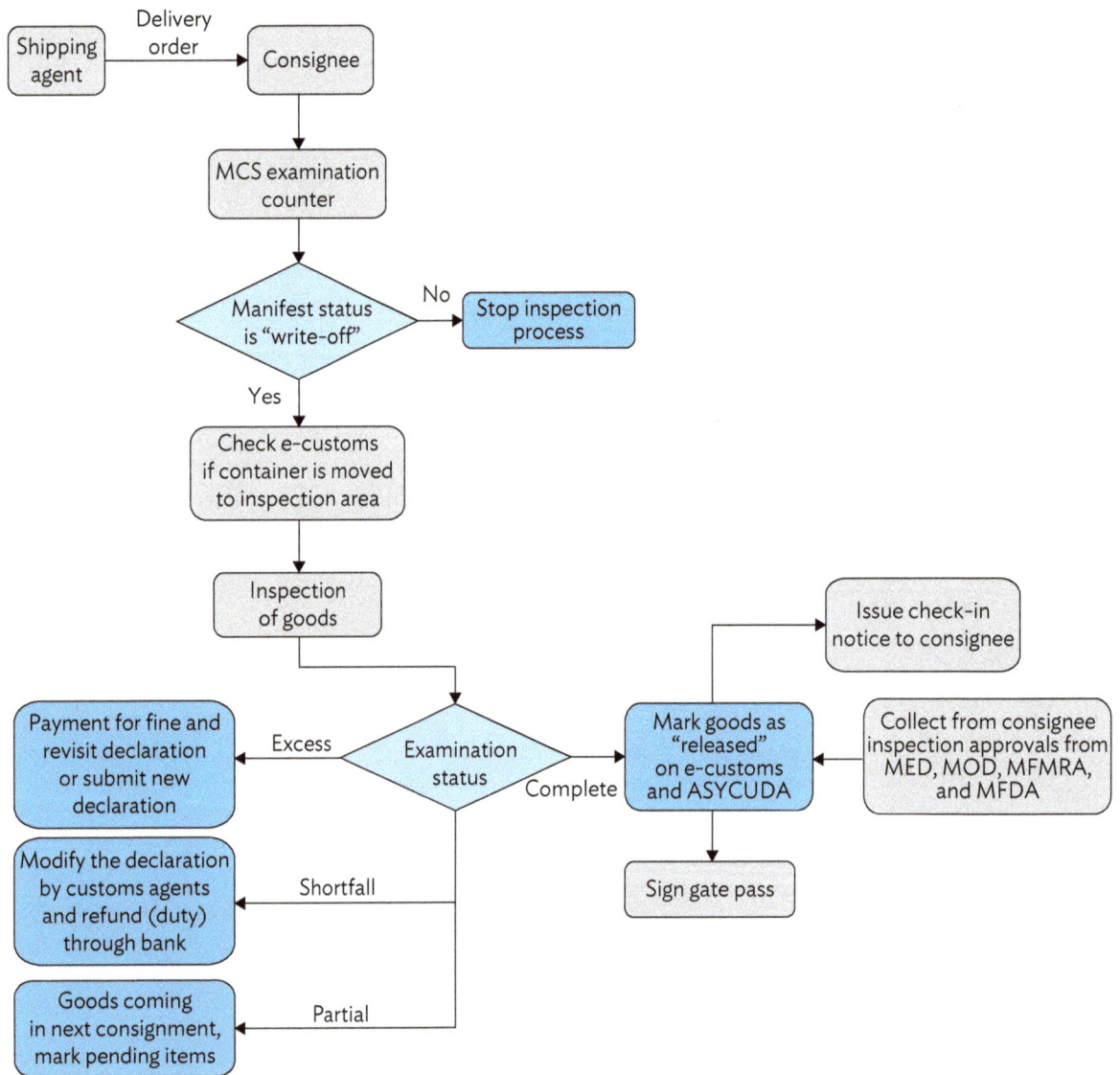

Shipping agent → (Delivery order) → Consignee

Consignee → MCS examination counter

MCS examination counter → Manifest status is "write-off"

Manifest status is "write-off" → No → Stop inspection process

Manifest status is "write-off" → Yes → Check e-customs if container is moved to inspection area

Check e-customs if container is moved to inspection area → Inspection of goods

Inspection of goods → Examination status

Examination status → Excess → Payment for fine and revisit declaration or submit new declaration

Examination status → Shortfall → Modify the declaration by customs agents and refund (duty) through bank

Examination status → Partial → Goods coming in next consignment, mark pending items

Examination status → Complete → Mark goods as "released" on e-customs and ASYCUDA

Mark goods as "released" on e-customs and ASYCUDA → Issue check-in notice to consignee

Collect from consignee inspection approvals from MED, MOD, MFMRA, and MFDA → Mark goods as "released" on e-customs and ASYCUDA

Mark goods as "released" on e-customs and ASYCUDA → Sign gate pass

ASYCUDA = Automated System for Customs Data; MCS = Maldives Customs Service; MED = Ministry of Economic Development; MFDA = Maldives Food and Drug Authority; MFMRA = Ministry of Fisheries, Marine Resources and Agriculture; MOD = Ministry of Defence.

Source: Author's analysis.

## Figure A.7: Maldives Port Trade Operations

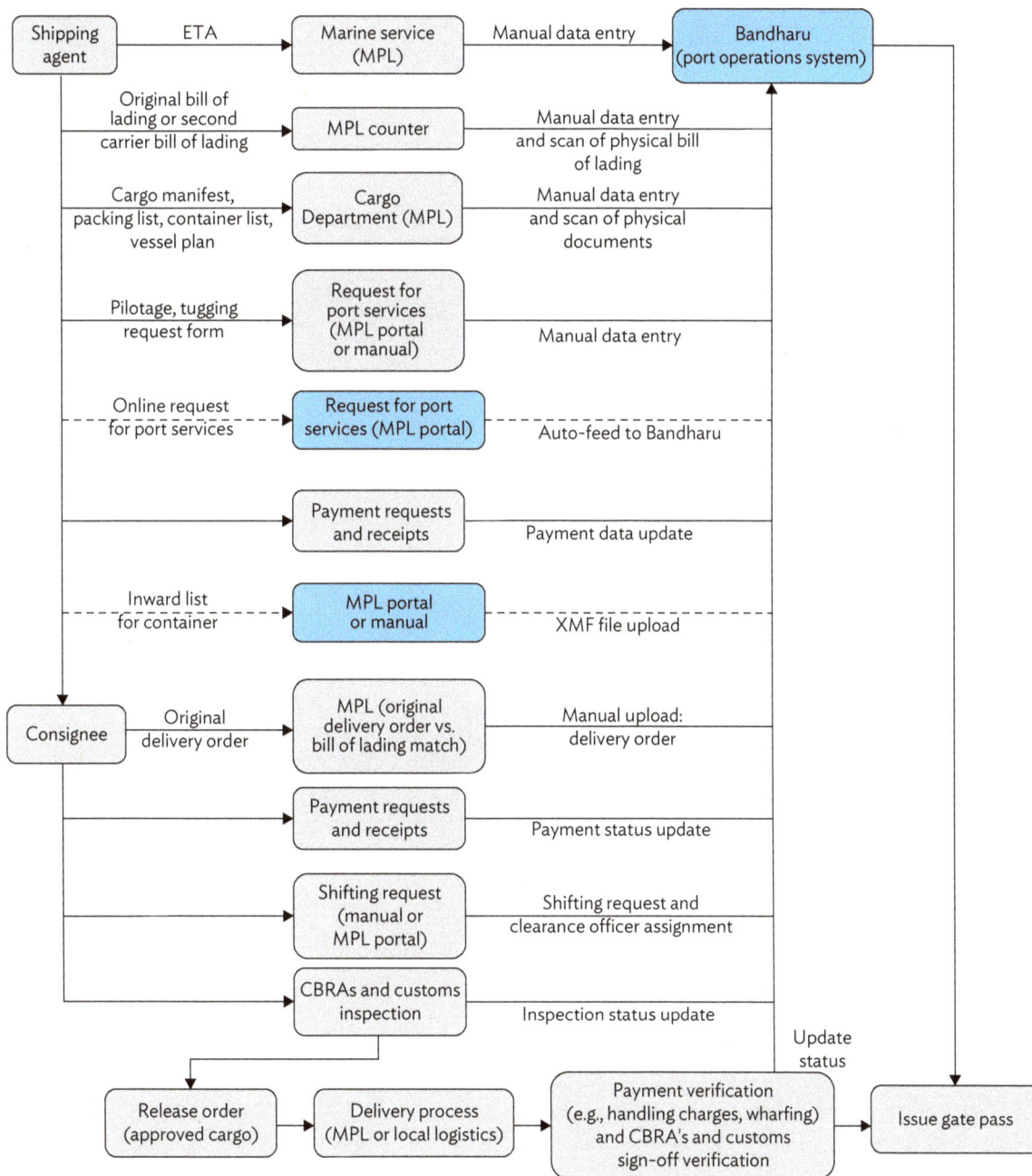

CBRA = cross-border regulatory agency, ETA = estimated time of arrival, MPL = Maldives Ports Limited.
Source: Author's analysis.

## Figure A.8: Air Cargo—Import and Export Trade Operations

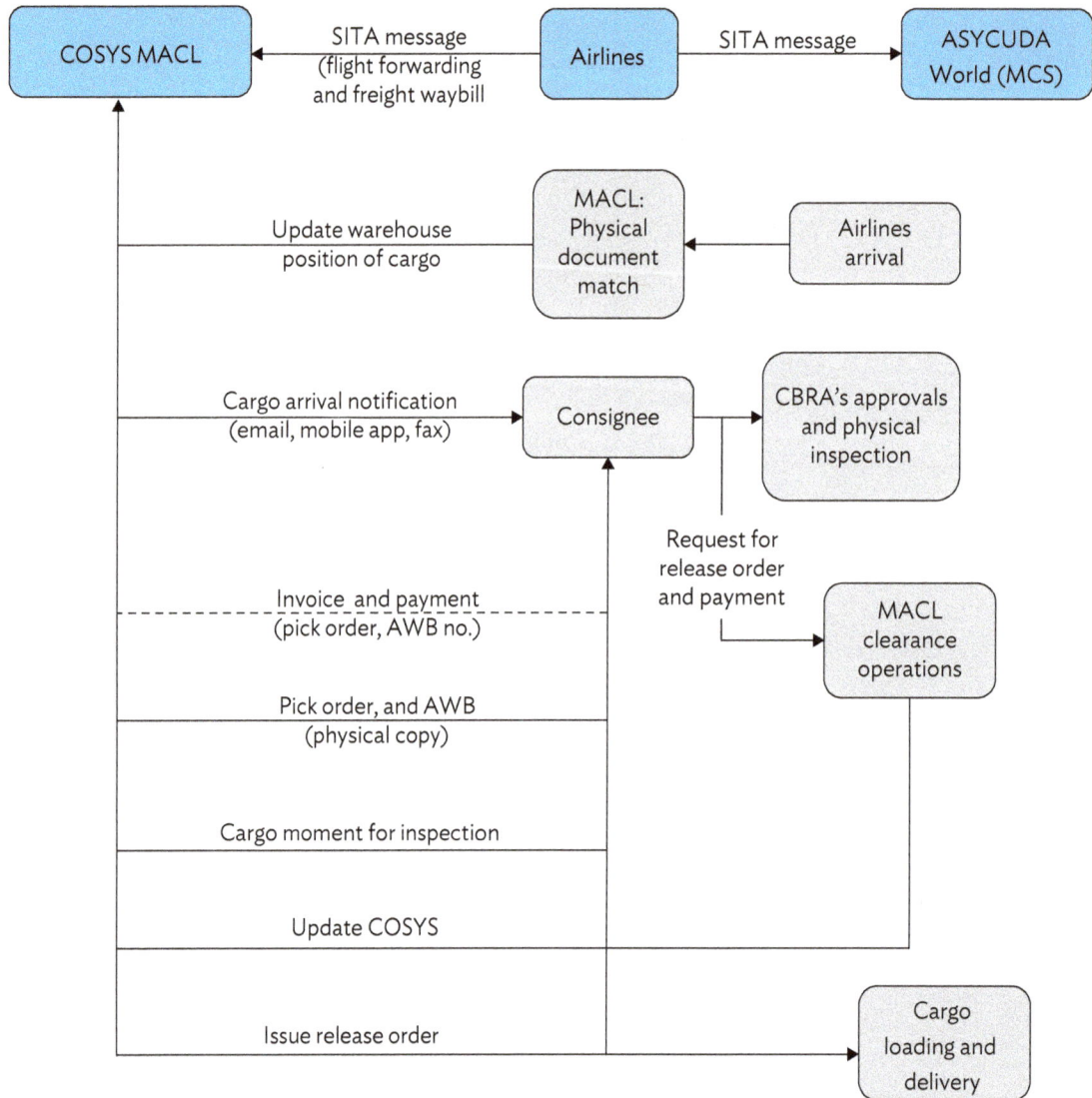

COSYS MACL

SITA message
(flight forwarding
and freight waybill

Airlines

SITA message

ASYCUDA
World (MCS)

Update warehouse
position of cargo

MACL:
Physical
document
match

Airlines
arrival

Cargo arrival notification
(email, mobile app, fax)

Consignee

CBRA's approvals
and physical
inspection

Request for
release order
and payment

Invoice  and payment
(pick order, AWB no.)

MACL
clearance
operations

Pick order, and AWB
(physical copy)

Cargo moment for inspection

Update COSYS

Issue release order

Cargo
loading and
delivery

ASYCUDA = Automated System for Customs Data, AWB = air waybill, CBRA = cross-border regulatory agency, COSYS = cargo operations system , MACL = Maldives Airports Company Limited, MCS = Maldives Customs Service, SITA = Société Internationale de Télécommunications Aéronautiques.

Note: Pick order is similar to a delivery order. The document allows the consignee to take ownership of the cargo.

Source: Author's analysis.

## Figure A.9: Export Trade

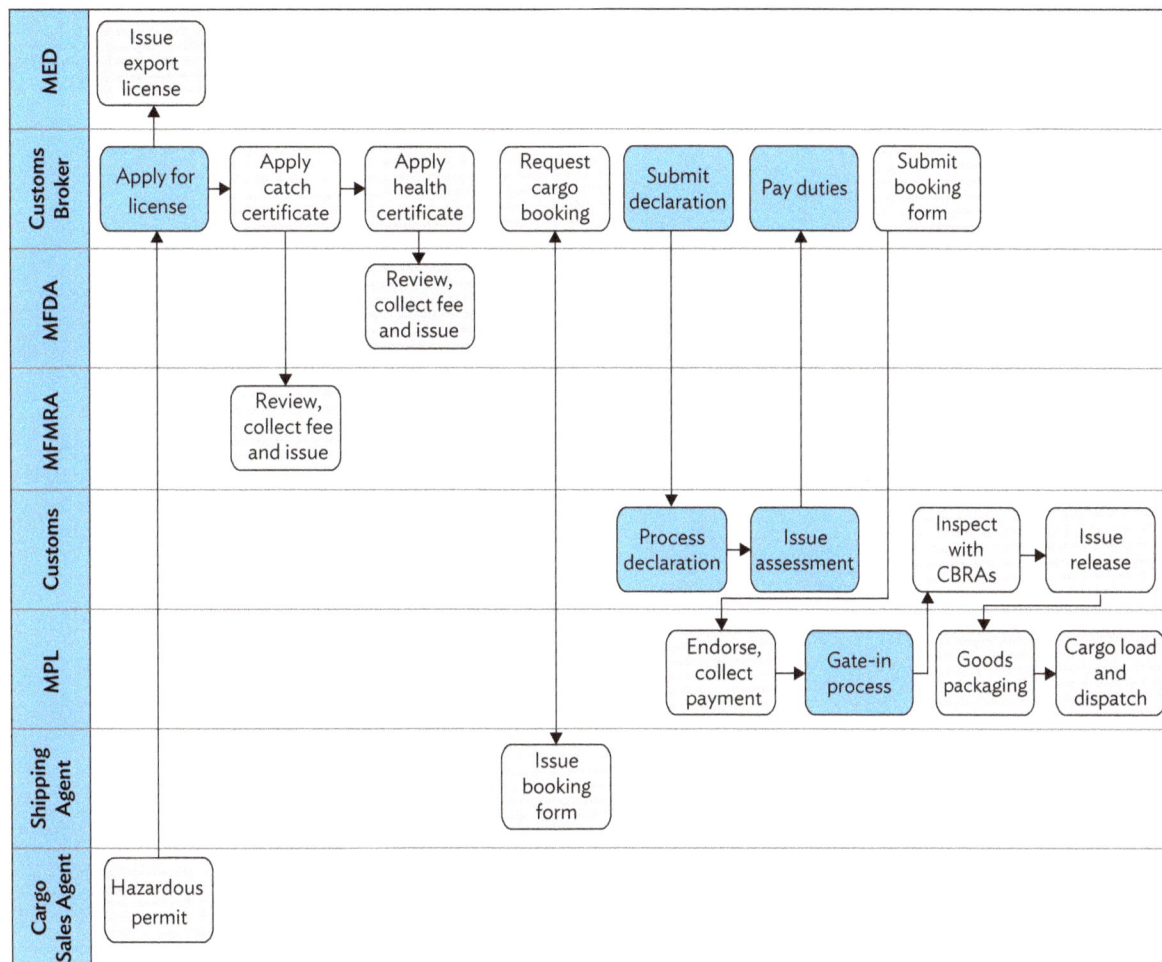

CBRA = cross-border regulatory agency; MED = Ministry of Economic Development; MFDA = Maldives Food and Drug Authority; MFMRA=Ministry of Fisheries; Marine Resource and Agriculture; MPL = Maldives Ports Limited.

Note: Gate-in process is performed by MPL to allow export goods to be brought into the inspection area.

Source: Author's analysis.

# Glossary

C#

C# (pronounced "see sharp") is a general-purpose, multi-paradigm programming language.

middleware

Software that acts as a bridge between an operating system or database and applications, especially on a network.

thick client

A thick client is a computer that does not require a connection to a server system to run (unlike a thin client). However, they can benefit from connecting to a network and a server.

thin client

In computer networking, this is a simple (low-performance) computer that has been optimized for establishing a remote connection with a server-based computing environment. The server does most of the work, which can include launching software programs, performing calculations, and storing data.